PROBLEM BOOK IN
THE THEORY OF FUNCTIONS

Volume I
Problems in the Elementary Theory of Functions

By DR. KONRAD KNOPP
Professor of Mathematics at the University of Tübingen

Translated by Dr. LIPMAN BERS

Associate Professor of Mathematics at Syracuse University

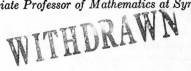
DOVER PUBLICATIONS, INC.
NEW YORK

Published in Canada by General Publishing Company, Ltd., 30 Lesmill Road, Don Mills, Toronto, Ontario.

Published in the United Kingdom by Constable and Company, Ltd., 10 Orange Street, London WC 2.

Problem Book in the Theory of Functions, Volume I, is a new English translation of the second edition of *Aufgabensammlung zur Funktionentheorie,* I Teil.

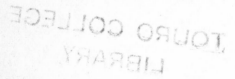
International Standard Book Number: 0-486-60158-7
Library of Congress Catalog Card Number: 49-7400

Manufactured in the United States of America
Dover Publications, Inc.
180 Varick Street
New York, N. Y. 10014

CONTENTS

FOREWORD

This translation follows the second edition of K. Knopp's *Aufgabensammlung zur Funktionentheorie*, I. Teil, except for a few minor changes.

The problems of the first five chapters concern material treated in the first volume of the same author's *Theory of Functions* (referred to as KI; the references are to the translation by Bagemihl, Dover Publications, Inc., 1945.) Familiarity with the first two chapters of C. Carathéodory's *Conformal Representation* (Cambridge University Press, 1932, referred to as C) is sufficient for the understanding of the problems in Chapter VI.

The solution of a problem often depends on that of a previous one. Some problems are more difficult than others; these are marked by asterisks. On many occasions, the reader will find carefully executed sketches helpful in the solution of problems. This applies particularly to the problems in Chapter VI.

The notations in this volume are as follows: Complex numbers (and points) are denoted by z_0, z_1, $\cdots$, w_0, w_1, $\cdots$ a, b $\cdots$, complex variables by z, ζ $\cdots$ w, w $\cdots$. (In §13, however, z_1, z_2 $\cdots$ denote variables.) Numbers conjugate to z, a, $\cdots$ are denoted by $\bar{z}$, $\bar{a}$ $\cdots$.

Real constants are denoted by x_0, x_1, $\cdots$, y_0, y_1, $\cdots$, u_0, $\cdots$, v_0, α, β, $\cdots$, λ, μ $\cdots$. We write $z = x + iy = r(\cos \varphi + i \cdot \sin \varphi)$, $x = \Re(z)$, $y = \Im(z)$, $r = |z|$, $\varphi = \text{am } z$.

Positive constants are denoted by r, ρ, δ, ϵ $\cdots$, positive integers by m, n, p $\cdots$.

Regions are denoted by capital German letters: $\mathfrak{G}$, $\mathfrak{M}$, $\cdots$, paths and curves by l. c. German and capital roman letters: $\mathfrak{s}$, $\mathfrak{p}$, $\cdots$, C. L, $\cdots$.

Part I—PROBLEMS

FUNDAMENTAL CONCEPTS

§1. Numbers and Points

(KI, 1–2)

1. Given a complex number $z_0 \neq 0$, find its reflection with respect to a) the origin, b) the real axis, c) the imaginary axis, d) the line $x - y = 0$, e) the line $x + y = 0$.

2. Show that $(1/2^{\frac{1}{2}})(|x| + |y|) \leq |z| \leq |x| + |y|$.

3. Find the loci of points z satisfying the following relations:

a) $|z| \leq 2$;   b) $|z| > 2$;   c) $\Re(z) \geq \dfrac{1}{2}$;

d) $0 \leq \Re(iz) < 2\pi$;   e) $\Re(z^2) = \alpha\left(\begin{smallmatrix}\geq\\=\\<\end{smallmatrix}0\right)$;

f) $\Im(z^2) = \alpha\left(\begin{smallmatrix}\geq\\=\\<\end{smallmatrix}0\right)$;   g) $|z^2 - z| \leq 1$;

h) $|z^2 - 1| = \alpha > 0$;   i) $\left|\dfrac{1}{z}\right| < \delta, \delta > 0$;

1

j) $\left| \dfrac{z-1}{z+1} \right| \leq 1;$ k) $\left| \dfrac{z-1}{z+1} \right| \geq 2;$

l) $\left| \dfrac{z}{z+1} \right| = \alpha > 0;$ m) $\left| \dfrac{z-z_1}{z-z_2} \right| = 1.$

4. When are z_1, z_2, z_3 collinear? (Consider the difference quotient $(z_1 - z_3)/z_2 - z_3).$)

5. When do z_1, z_2, z_3, z_4 lie on a circle or on a straight line? (Consider the cross-ratio $(z_1 - z_3)/(z_2 - z_3) \div (z_1 - z_4)/(z_2 - z_4).$)

6. What is the geometrical meaning of the identity $|z_1 + z_2|^2 + |z_1 - z_2|^2 = 2(|z_1|^2 + |z_2|^2)$?

7. Find the point z dividing the segment $z_1 \cdots z_2$ in the ratio $\lambda_1 \div \lambda_2$ $(\lambda_1 + \lambda_2 \neq 0).$

8. Find the mass center of the triangle z_1, z_2, z_3 when a) each vertex z_i carries the same mass λ, b) the vertices carry the masses λ_1, λ_2, λ_3. c) Show that the mass center found in b) lies within the triangle if all three masses are positive.

9. The masses λ_1, λ_2, $\cdots$, λ_k are situated at z_1, z_2, $\cdots$, z_k. Show that the mass center of this system is $z = (\lambda_1 z_1 + \lambda_2 z_2 + \cdots + \lambda_k z_k)/(\lambda_1 + \lambda_2 + \cdots + \lambda_k).$

10. Given that $z_1 + z_2 + z_3 = 0$, $|z_1| = |z_2| = |z_3| = 1$, show that z_1, z_2, z_3 are the vertices of an equilateral triangle inscribed into the unit circle.

11. Given that $z_1 + z_2 + z_3 + z_4 = 0$, $|z_1| = |z_2| = |z_3| = |z_4| = 1$, show that z_1, z_2, z_3, z_4 are the vertices of a rectangle inscribed into the unit circle.

12. When are two triangles, z_1, z_2, z_3 and z_1', z_2', z_3' similar and similarly situated? (Cf. problem 4.)

*13. a) Given two points z_1, z_2, $|z_1| < 1$, $|z_2| < 1$, show that for every point $z \neq 1$ belonging to the triangle z_1, z_2, 1

$$\frac{|1 - z|}{1 - |z|} \leq K$$

where $K = K(z_1, z_2)$ is a constant depending only on z_1 and z_2.

b) Determine the smallest value of K for $z_1 = (1 + i)/2$, $z_2 = (1 - i)/2$.

§2. Point Sets. Paths. Regions

(KI, 3–4)

1. Show that the set of roots of algebraic equations of the form

$$a_0 z^n + a_1 z^{n-1} + \cdots + a_{n-1} z + a_n = 0,$$

the $a_n's$ being Gaussian integers, is countable. (A set is countable if its elements can be arranged in a sequence. A complex number z is called a Gaussian integer if $\Re(z)$ and $\Im(z)$ are real integers.)

2. Show that the set of all numbers $z = x + iy$, x and y being rational numbers, is countable.

3. Order the set of all numbers $z = 1/m + i/n$ (m, n positive integers) into a sequence.

4. Find the greatest lower bound α, the least upper bound β, the lower limit λ, and the upper limit μ of the following real sets. (Indicate whether or not α, β, λ, μ belong to the set considered).

a) The set of rational numbers p/q with even q and $p^2/q^2 \leq 10$.

b) The set of numbers of the form $(1 \pm 1/n)^n$.

c) The set of numbers of the form $(1 \pm 1/n^2)^n$.

d) The set of numbers of the form $n \pm 1/n$.

e) The set of numbers of the form $n \pm 1/3$.

f) The set of numbers of the form $1/m + 1/n$.

g) The set of numbers of the form $(1/m + 1/n)^{m+n}$.

h) The set of numbers of the form $\pm 1/m \pm 1/n$.

i) The set of numbers of the form $1 + (-1)^n + (-1)^n/n$.

k) The set of all numbers which may be written as infinite decimal fractions of the form $.\alpha_1\alpha_2\alpha_3 \cdots$, α_i odd.

(In b) to i) n and m denote arbitrary positive integers.)

*5. Show that each point of the set defined in 4k) is a limit point of the set.

6. Show that $\alpha = \lambda$ whenever α does not belong to the set and $\beta = \mu$ whenever β does not belong to the set.

7. Is the set defined by the relation $|z| + \Re(z) \leq 1$ bounded? What domain does it occupy?

8. Find all limit points of the following sets:

a) $1/m + i/n$, m and n positive integers,

b) $|z| < 1$,

c) $|z| > 1$,

d) the set defined in problem 2,

e) the set of all non-real z in the domain interior to the unit circle.

*9. Is the set defined in problem 4k) closed?

10. Is a limit point of a point set which does not belong to the set a boundary point of the set?

11. Show that a boundary point of a point set M which belongs to M is a limit point of the complementary set M'. (M' consists of all points which do not belong to M.)

12. Show that the set of all boundary points of a point set M is closed.

*13. Given two disjunct *closed* point sets M' and M'', one of which, say M', is bounded, show that there exists a positive number d such that $|z' - z''| \geq d$ whenever z' belongs to M' and z'' to M''. Show that among all such numbers d there exists a largest number d_0.

14. Show that an arc of the continuous curve

$$y = \begin{cases} x \sin(\pi/x), & x \neq 0 \\ 0 \text{ for } x = 0 \end{cases}$$

containing the origin is not rectifiable.

15. Let $\mathfrak{M}$ consist of all points of the upper half-plane $[\Im(z) > 0]$ except those lying on the segments $z = it, z = \pm 1/n + it, n = 1, 2, 3, \cdots, 0 < t \leq 1$. Is $\mathfrak{M}$ a region? Find the boundary points of $\mathfrak{M}$. Is $i/2$ a boundary point? Does there exist a path leading from $z_0 = 2 + i$ to $i/2$ and situated (except for the endpoint $i/2$) within $\mathfrak{M}$?

16. Consider the spiral S defined by

$$z = z(t) = \begin{cases} e^{(-1+i)/t}, & 0 < t \leq 1, \\ 0 \text{ for } t = 0. \end{cases}$$

Is S a path leading from $z_1 = z(1)$ to $z_0 = 0$?

*17. Let $\mathfrak{G}$ be a plane region, $\mathfrak{G}_1$ its image under stereographic projection, M the set of boundary points of $\mathfrak{G}_1$. Show that $\mathfrak{G}$ is simply connected if and only if M is connected. (A closed set is called connected if it can not be divided into two closed sub-sets without a common element.)

18. Is the region defined in problem 15 simply connected?

19. Show that a simply connected region $\mathfrak{M}$ on the surface of a sphere which does not contain two points of the sphere does not contain infinitely many points of the sphere.

CHAPTER II

INFINITE SEQUENCES AND SERIES

§3. Limits of Sequences. Infinite Series with Constant Terms

(KI, 2–3)

1. Let ζ be a limit point of the sequence z_1, z_2, $\cdots$, z_n, $\cdots$. Show that the sequence contains a subsequence z_1', z_2', $\cdots$ which converges to ζ.

2. If $z_n \to \zeta$, then

$$z_n' = \frac{z_1 + z_2 + \cdots + z_n}{n} \to \zeta.$$

Is this true if $\zeta = \infty$?

3. If $z_n \to \zeta$, then

$$z_n' = \frac{p_1 z_1 + p_2 z_2 + \cdots + p_n z_n}{p_1 + p_2 + \cdots + p_n}$$

$$\equiv \frac{P_1 z_1 + (P_2 - P_1)z_2 + \cdots + (P_n - P_{n-1})z_n}{P_n} \to \zeta$$

where p_1, p_2, $\cdots$ is any sequence of positive numbers such that $P_n = (p_1 + p_2 + \cdots + p_n) \to +\infty$.

4. If $z_n \to \zeta$, then

$$z_n' = \frac{b_1 z_1 + b_2 z_2 + \cdots + b_n z_n}{b_1 + b_2 + \cdots + b_n}$$

$$\equiv \frac{B_1 z_1 + (B_2 - B_1)z_2 + \cdots + (B_n - B_{n-1})z_n}{B_n} \to \zeta$$

7

if b_1, b_2, $\cdots$ are complex numbers such that for all n the numbers $\beta_n = (|\, b_1 + b_2 + \cdots + b_n\,|)/(|\, b_1\,| + |\, b_2\,| + \cdots + |\, b_n\,|)$ exceed some fixed positive number β, and such that $(|\, b_1\,| + |\, b_2\,| + \cdots + |\, b_n\,|) \to +\infty$.

5. Let there be given infinitely many numbers $a_{\kappa\lambda}$ arranged in the form

$$
\begin{array}{llll}
a_{11} & & & \\
a_{21} & a_{22} & & \\
a_{31} & a_{32} & a_{33} & \\
a_{41} & a_{42} & a_{43} & a_{44}
\end{array}
$$

. . . .

and satisfying the conditions: 1) for every fixed p, $a_{np} \to 0$, 2) there exists a positive constant M such that $|\, a_{n1}\,| + |\, a_{n2}\,| + \cdots + |\, a_{nn}\,| \leq M$ for all n. Show that if $z_n \to 0$, then

$$
z'_n = a_{n1}z_1 + a_{n2}z_2 + \cdots + a_{nn}z_n \to 0.
$$

6. Assume that in addition to conditions 1) and 2) of the preceding problem the numbers $a_{\kappa\lambda}$ also satisfy the condition 3) $A_n = a_{n1} + a_{n2} + \cdots + a_{nn} \to 1$. Show that if $z_n \to \zeta$, then

$$
z'_n = a_{n1}z_n + a_{n2}z_n + \cdots + a_{nn}z_n \to \zeta.
$$

Show that this theorem contains as special cases the theorems stated in problems 2, 3, 4.

7. a) If $z'_n \to \zeta'$ and $z''_n \to \zeta''$, then

$$
z_n = \frac{z'_1 z''_1 + z'_2 z''_2 + \cdots + z'_n z''_n}{n} \to \zeta'\zeta'',
$$

*b) and

$$z_n = \frac{z_1' z_n'' + z_2' z_{n-1}'' + \cdots + z_n' z_1''}{n} \to \zeta' \zeta'',$$

*c) and

$$z_n = a_{n1} z_1' z_n'' + a_{n2} z_2' z_{n-1}'' + \cdots + a_{nn} z_n' z_1'' \to \zeta' \zeta''$$

provided the numbers $a_{\kappa\lambda}$ satisfy the conditions 1), 2), 3) of problems 5 and 6, as well as 4) each "diagonal" sequence converges to 0, i.e. for each fixed p, $a_{n,n-p+1} \to 0$.

8. If $z_n \to \zeta$, then

$$z_n' = \frac{\binom{n}{1} z_1 + \binom{n}{2} z_2 + \cdots + \binom{n}{n} z_n}{2^n} \to \zeta.$$

9. If $z_n' \to 0$, $z_n'' \to 0$, and if there exists a constant M such that $|z_1''| + |z_2''| + \cdots + |z_n''| \leq M$ for all n, then

$$z_n = z_1' z_n'' + z_2' z_{n-1}'' + \cdots + z_n' z_1'' \to 0.$$

10. The following condition is necessary and sufficient in order that the infinite series $\sum_{n=0}^{\infty} c_n$ be convergent: for every sequence of positive integers p_1, p_2, $\cdots$, p_n, $\cdots$,

$$T_n = (c_{n+1} + c_{n+2} + \cdots + c_{n+p_n}) \to 0.$$

11. Given two sequences of numbers a_0, a_1, a_2, $\cdots$ and b_0, b_1, b_2, $\cdots$, set $a_0 + a_1 + \cdots + a_n = s_n$. For $n \geq 0$ and $p \geq 1$, show that

$$\sum_{\nu=n+1}^{n+p} a_\nu b_\nu = \sum_{\nu=n+1}^{n+p} s_\nu (b_\nu - b_{\nu+1}) - s_n b_{n+1} + s_{n+p} b_{n+p+1}.$$

(Abel's summation by parts.)

12. Let $\sum_{n=0}^{\infty} a_n$ be a (convergent or divergent) infinite series. Set $a_0 + a_1 + \cdots + a_n = s_n$. Let b_0 , b_1 , $\cdots$ be a sequence of numbers such that

   1. the sequence of numbers $s_n b_{n+1}$ converges,

   2. the series $\sum_{n=0}^{\infty} s_n(b_n - b_{n+1})$ converges.

Show that the series $\sum_{n=0}^{\infty} a_n b_n$ converges.

13. Show that the conditions of the preceding theorem are satisfied in each of the following three cases:

a) $b_{n-1} > b_n$ for all n, $b_n \to 0$, (s_n) is bounded;

b) $b_{n-1} > b_n > \alpha$ for all n, Σa_n converges;

c) $b_n \to 0$, $\Sigma \mid b_n - b_{n+1} \mid$ converges, (s_n) is bounded.

*14. Let the partial sums s_n of the series Σa_n be such that the sequence $(s_n/n^{1/2})$ is bounded. Let the positive numbers b_n be such that $b_{n-1} > b_n \to 0$, $n^{1/2} b_n \to 0$, and $\Sigma n^{1/2}(b_n - b_{n+1})$ converges. Show that under these conditions $\Sigma a_n b_n$ converges.

§4. Convergence Properties of Power Series

(KI, 17–20)

1. Find the radii of convergence of the power series $\sum_{n=0}^{\infty} a_n z^n$ with

a) $a_n = \dfrac{1}{n}, \; = n, \; = \dfrac{1}{n^p}, \; = \dbinom{n + \alpha}{n}$;

b) $a_n = \dfrac{1}{n^n}, \; = n^{\log n}, \; = (\log n)^n$;

c) $a_n = \dfrac{n!}{n^n}, \; = \dfrac{(n!)^2}{n^n}, \; = \left(1 - \dfrac{1}{n}\right)^{n^2}, \; = n^{(\log n)^2}$;

d) $a_n = \tau(n) =$ the number of factors of n,

$\qquad = \phi(n) =$ the number of integers $\leqq n$ which

$\qquad$ are prime to n;

e) $\begin{cases} a_{2k} \\ a_{2k+1} \end{cases} = \begin{cases} 2^k \\ 3^k, \end{cases} = \begin{cases} k^2 \\ k^{1/2}, \end{cases} = \begin{cases} 1/k! \\ (\log k)^k. \end{cases}$

2. Let there be given a power series $\sum_{n=0}^{\infty} a_n z^n$. If for some value z_0 and for two positive numbers K and k, either a) $|a_n z_0|^n < Kn^k$, or b) $|a_0 + a_1 z_0 + \cdots + a_n z_0^n| < Kn^k$, for all n, then the power series converges absolutely for every z such that $|z| < |z_0|$.

3. Let the series $\Sigma a_n z^n$ have the radius of convergence r and the series $\Sigma a_n' z^n$ the radius of convergence r'. Find the radii of convergence R of the following series: $\Sigma(a_n + a_n')z^n$, $\Sigma a_n a_n' z^n$, $\Sigma(a_n/a_n')z^n$. (In the last case it is assumed that no a_n' vanishes.)

4. Given that the series $\Sigma a_n z^n$ has the radius of convergence $r > 0$. Find the radii of convergence of the following series: $\Sigma a_n n^p z^n$, $\Sigma(a^n/n^p)z^n$, $\Sigma(a_n/n!)z^n$, $\Sigma a_n n! z^n$.

*5. At which points on the circle of convergence does the power series

$$\sum_{n=1}^{\infty} \frac{z^n}{n}$$

converge?

6. If $r > 0$ is the radius of convergence of the series $\Sigma a_n z^n$, and if at a point z_0, $|z_0| = r$, the series converges absolutely, show that $\Sigma a_n z^n$ converges absolutely and uniformly for $|z| \leq r$.

7. Consider the power series

$$z^p + \frac{z^{2p}}{2} + \frac{z^{3p}}{3} + \cdots,$$

p being a positive integer. At which points on the circle of convergence does the series converge?

8. If the coefficients of the power series $\Sigma a_n z^n$ are real numbers and $a_n > a_{n+1}$, $a_n \to 0$, then a) the series has a radius of convergence $r \geq 1$, and b) if $r = 1$, then the series converges at all points on the circle of convergence, except perhaps the point $z = 1$.

9. The assertion of the preceding problem holds also for series with complex coefficients a_n, provided that $a_n \to 0$ and $\Sigma \mid a_n - a_{n+1} \mid$ converges.

10. How must one modify the statements in problems 8b and 9 if $r > 1$?

11. Let r be the radius of convergence of $\Sigma a_n z^n$, ρ that of $\Sigma b_n z^n$. Set $c_n = a_0 b_n + a_1 b_{n-1} + \cdots + a_n b_0$. What is the radius of convergence R of the series $\Sigma c_n z^n$?

*12. Let there be given the power series

$$w = f(z) = a_0 + a_1(z - z_0) + a_2(z - z_0)^2 + \cdots$$

with radius of convergence $r > 0$, and the power series

$$W = g(w) = A_0 + A_1(w - w_0) + A_2(w - w_0)^2 + \cdots$$

with radius of convergence $R > 0$. Using the relation

$$w - w_0 = (a_0 - w_0) + \sum_{n=1}^{\infty} a_n(z - z_0)^n,$$

obtain series for $(w - w_0)^k$, $k = 2, 3, \cdots$. Substitute these expressions into the series for W to obtain the power series

$$W = \sum_{n=0}^{\infty} b_n (z - z_0)^n.$$

For which values of z is this series certainly convergent? For which values is it certainly divergent? If the series converges, is it true that $W = g(f(z))$?

*13. If $w = f(z) = a_0 + a_1 z + a_2 z^2 + \cdots$ and $a_0 \neq 0$, then $W = 1/w = 1/(f(z))$ can be expanded in a power series $b_0 + b_1 z + b_2 z^2 + \cdots$. Find recursion formulas for the coefficients b_n and a lower bound for the radius of convergence of $\Sigma b_n z^n$.

14. Form a power series whose radius of convergence is 1 and which diverges at p preassigned points of the unit circle, while converging at all other points on this circle.

**15. Does there exist a power series with the following properties: the radius of convergence is 1, the series converges at $z = 1$ and diverges at all other points on the unit circle?

FUNCTIONS OF A COMPLEX VARIABLE

§5. Limits of Functions. Continuity and Differentiability

(KI, 5–7)

1. Investigate the following two functions:

(a) $f(z) = 0$ whenever $|z|$ is an irrational number; $f(0) = 0$; $f(z) = 1/q$ whenever $|z|$ is a rational number, $|z| = p/q$, p and q being positive relatively prime integers.

(b) $f(0) = 0$; $f(z) = \sin\theta$ for $z = r(\cos\theta + i\sin\theta)$, $r > 0$. At which points of the plane are these functions continuous? At which points are they discontinuous?

2. If $f(z)$ is continuous at ζ and $z_n \to \zeta$, then $w_n = f(z_n) \to f(\zeta)$.

3. Conversely, if $f(z)$ is defined at ζ and in a neighborhood of this point and if for every sequence of points z_n of this neighborhood $w_n = f(z_n) \to f(\zeta)$ whenever $z_n \to \zeta$, then $f(z)$ is continuous at ζ.

4. Is the function $f(z) = 1/(1 - z)$ continuous in the domain interior to the unit circle? Is it uniformly continuous in this domain?

5. Set $f(z) = e^{-1/|z|}$ for $z \neq 0$. Is this function continuous (uniformly continuous) in the domain $0 < |z| < 1$?

6. For $z \neq 0$ set

$$f_1(z) = \frac{\Re(z)}{1 + |z|}, \quad f_2(z) = \frac{\Re(z)}{|z|}, \quad f_3(z) = \frac{[\Re(z)]^2}{|z|},$$

$$f_4(z) = \frac{\Re(z^2)}{|z^2|}, \quad f_5(z) = \frac{[\Re(z^2)]^2}{|z^2|}.$$

Assign to all five functions the value 0 at $z = 0$. Which of these functions are continuous at $z = 0$?

7. Replace in the definition of the above function $\Re(z)$ by $\Im(z)$ and answer the same question.

*8. Let $f(z)$ be defined for $|z| < 1$ (*not* for $|z| \leq 1$) and be uniformly continuous in this domain. For every sequence of points z_n, $|z_n| < 1$, converging toward a point ζ of the unit circle, $\lim_{n\to\infty} f(z_n)$ exists and depends only on ζ. (In other words: $f(z)$ possesses uniquely determined boundary values.)

*9. The boundary values of the function $f(z)$ in problem 8 are continuous.

10. At which points are the functions of problems 1 and 5 differentiable? (Give detailed proofs for each function.)

11. If two real-valued functions, $\phi(x, y)$ and $\psi(x, y)$ defined in a domain G possess continuous partial derivatives satisfying the Cauchy-Riemann equations, the same is true for the following pairs of functions: $\phi_1 = \phi^2 - \psi^2$, $\psi_1 = 2\phi\psi$; $\phi_2 = e^\phi \cos \psi$, $\psi_2 = e^\phi \sin \psi$.

12. In a domain $\mathfrak{G} : f'(z) \equiv 0$. Without separating f into its real and imaginary parts, prove that f is constant in $\mathfrak{G}$.

*13. Let $f'(z)$ exist and be continuous in $|z - \zeta| < \rho$. Let (z_n) and (z_n') be two sequences of points from this domain; $z_n \neq z_n'$ for all n, $z_n \to \zeta$, $z_n' \to \zeta$. Prove that

$$\frac{f(z_n') - f(z_n)}{z_n' - z_n} \to f'(\zeta).$$

(Hint: Separate the difference quotient into real and imaginary parts and apply the mean-value theorem.)

*14. If $f(z)$ possesses a continuous derivative in $|z - \zeta| < \rho$, then $f(z)$ is *uniformly* differentiable in $|z - \zeta| \leq \rho' < \rho$; i.e., for every $\epsilon > 0$ there exists a $\delta < 0$ such that

$$\left| \frac{f(z') - f(z)}{z' - z} - f'(z) \right| < \epsilon$$

whenever $|z - \zeta| < \rho'$, $|z' - \zeta| < \rho'$, and $0 < |z' - z| < \delta$.

§6. Simple Properties of the Elementary Functions

(KI, 23–26)

1. For $0 < |z| < 1$, $(1/4)|z| < |e^z - 1| < (7/4)|z|$.

2. $|e^z - 1| \leq e^{|z|} - 1 \leq |z| e^{|z|}$.

3. If x is real and different from 0, then

(a) $e^x > 1 + x$,

(b) $e^x < 1/(1 - x)$ for $x < 1$,

(c) $x/(1 + x) < 1 - e^{-x} < x$ for $x > -1$,

(d) $x < e^x - 1 < x/(1 - x)$ for $x < 1$,

(e) $1 + x > e^{x/(1+x)}$ for $x > -1$,

(f) $e^x > x^p/p!$ for $x > 0$ and all positive integers p,

(g) $e^x > (1 + x/y)^y > e^{xy/(x+y)}$ for $x > 0$, $y > 0$,

(h) $e^x < 1 - (x/2)$ for $0 < x < 1$.

4. $(1 + z/n)^n \to e^z$ for all z.

5. If the sequence (z_n) converges to ζ, then $(1 + z_n/n)^n \to e^\zeta$.

6. Show that the equation $e^z = 1$ has no solutions other than $z = 2k\pi i$ ($k = 0, \pm 1, \pm 2, \cdots$).

7. By direct multiplication of power series show that

$$e^{z_1} \cdot e^{z_2} = e^{z_1 + z_2}.$$

8. Using the power series defining e^z prove that this function is periodic.

9. Let z approach infinity along a ray through the origin (am $z =$ const.). Find all directions for which

$$\lim e^z$$

exists.

10. Let z move as in the preceding problem. How does $z + e^z$ behave?

11. z approaches infinity along the hyperbola $xy = 1$. Find the four values of lim e^z corresponding to the four branches of the hyperbola.

12. Does lim e^z exist when z approaches infinity along one of the branches of the parabola $y = x^2$?

13. Prove (by direct multiplication of the power series) that

$$\cos (z_1 + z_2) = \cos z_1 \cdot \cos z_2 - \sin z_1 \cdot \sin z_2 ,$$

$$\sin (z_1 + z_2) = \sin z_1 \cdot \cos z_2 + \cos z_1 \cdot \sin z_2 ,$$

$$\cos^2 z + \sin^2 z = 1.$$

14. Along which curves in the z-plane are the functions e^z, $\cos z$, $\sin z$ real?

15. What is the most convenient way to compute e^z, $\cos z$, $\sin z$? Compute e^{2+i}, $\cos (5 - i)$, $\sin (1 - 5i)$ to three places after the decimal point.

16. Find all solutions of the following equations

$$\sin z = 1000, \sin z = 5i, \sin z = 1 - i,$$

$$\cos z = 2i, \cos z = 4 + 3i, \cos z = 5.$$

17. Set

$$\tan z = \frac{\sin z}{\cos z} = a_0 + a_1 z + a_2 z^2 + a_3 z^3$$
$$+ a_4 z^4 + a_5 z^5 + \cdots .$$

Find the coefficients a_0, a_1, $\cdots$, a_5 by division of the power series for $\sin z$ and $\cos z$.

18. Is the equation $\cos z = (1 - \sin^2 z)^{1/2}$ correct for all values of z?

19. Prove that for the principal value of the logarithm

$$| \log z | \le \frac{\rho}{1 - \rho}$$

if $| z - 1 | \le \rho < 1$.

*20. Give a rigorous proof of the formula

$$\frac{d \arc \sin z}{dz} = \frac{1}{(1 - z^2)^{1/2}}$$

taking in account the fact that both sides of this equation contain multiple-valued functions.

21. Find a relation between

(a) arc sin z and log z,

(b) arc tan z and log z.

22. What is the meaning of the symbol i^i? What is the meaning of a^b (a, b may be complex, $a \ne 0$)?

23. Is the formula $dz^a = az^{a-1} dz$ correct for a complex a? Which determinations of the multiple-valued functions must be used?

24. The functions

$$\text{sh}z = \frac{e^z - e^{-z}}{2}, \qquad \text{ch}z = \frac{e^z + e^{-z}}{2}$$

have many properties similar to those of $\sin z$ and $\cos z$. Show that

(a) $\text{ch}^2 z - \text{sh}^2 z = 1$,

(b) $(\text{sh}z)' = \text{ch}z, (\text{ch}z)' = \text{sh}z$,

(c) both functions possess the primitive period $2\pi i$,

(d) $\text{sh}(z_1 + z_2) = \text{sh}z_1\text{ch}z_2 + \text{ch}z_1\text{sh}z_2$,

$\text{ch}(z_1 + z_2) = \text{ch}z_1\text{ch}z_2 + \text{sh}z_1\text{sh}z_2$.

25. Along which curves are the functions $\text{sh}z$ and $\text{ch}z$ real?

CHAPTER IV

INTEGRAL THEOREMS

§7. Integration in the Complex Domain

(KI, 8–11)

1. Compute $\int_{-i}^{+i} |z| \, dz$ along a) a straight segment, b) the left half of the unit circle, c) the right half of the unit circle.

2. Compute $\int_L \Re(z) \, dz$, where (L) denotes
a) the unit circle traversed once in the positive direction from $+1$ to $+1$,
b) the straight segment from z_1 to z_2,
c) the circle $|z - z_0| = r$ traversed once in the positive direction.

3. Compute (without use of Cauchy's theorem) the integral

$$\int_L (z - z_0)^m \, dz \qquad \left(m \text{ an integer} \gtreqless 0 \right)$$

where
a) L is a square with center at z_0 and with sides parallel to the coordinate axes,
b) L is an ellipse with center at z_0 and axes parallel to the coordinate axes. (In this case assume that either m is even or $m = -1$).

4. Let $z^{1/2}$ denote the principal value of the square root. Compute the integral

$$\int_{+1}^{-1} \frac{dz}{z^{1/2}}$$

a) along the upper half of the unit circle,
b) along the lower half.

*5. Compute the integral

$$\int_L \frac{e^{iz}}{z}\, dz$$

for the following paths:

L_1 : straight segment from $\rho > 0$ to $r > \rho$,

L_2 : arc of $|z| = r$ from $+r$ to $-r$ through the upper half-plane,

L_3 : straight segment from $-r$ to $-\rho$,

L_4 : arc of $|z| = \rho$ from $-\rho$ to $+\rho$ through the upper half-plane.

*6. Let the value of the integral in problem 5 taken along L_i be denoted by $I_,$.

a) Find $\lim I_2$ as $r \to +\infty$.

b) Find $\lim I_4$ as $\rho \to 0$.

Hint: Add $i \int_0^\pi 1 \cdot dt = \pi i$ to I_4 and show that the new integral $\to 0$.

7. Let $f(z)$ be continuous for $|z - z_0| > r_0$. Let $\mathfrak{k}_r$ denote the circle $|z - z_0| = r$ or an arc of this circle. Let $M(r)$ denote the maximum of $f(z)$ for $|z - z_0| = r_0$. If $rM(r) \to 0$ as $r \to +\infty$, then

$$\lim_{r \to +\infty} \int_{\mathfrak{k}_r} f(z)\, dz = 0.$$

8. Let $f(z)$ be continuous for $0 < |z - z_0| = r_0$, but not necessarily at z_0 . If $rM(r) \to 0$ as $r \to 0$, then

$$\lim_{r \to 0} \int_{\mathfrak{k}_r} f(z)\, dz = 0.$$

(The notations are the same as in problem 7.)

9. Let L denote the radius of the unit circle forming the angle α with the positive real axis. For which values of α is the integral

$$\int_L e^{-1/z} \, dz$$

defined?

10. Let L and α have the same meaning. For which values α are the integrals

a) $\int_L e^{-1/z^2} \, dz$,   b) $\int_L e^{-1/z^p} \, dz$ (p positive integer)

defined?

11. How should one define the convergence of the (improper) integral

$$\int_{aL}^{b} f(z) \, dz$$

when $f(z)$ is continuous on L except at b and $f(z)$ does not necessarily remain bounded as $z \to b$ along L?

12. Let $f(z)$ be continuous along the path L leading from a to ∞. How should one define the convergence of the (improper) integral

$$\int_{aL}^{\infty} f(z) \, dz \ ?$$

§8. Cauchy's Integral Theorems and Integral Formulas

(KI, 12–16)

1. Determine the integrals in §7, problem 3 using Cauchy's integral theorem.

2. Using Cauchy's integral theorem, find

$$\int_C \frac{dz}{1 + z^2}$$

where C is the circumference

a) $|z - i| = 1$,   b) $|z + i| = 1$,   c) $|z| = 2$

traversed in the positive direction.

*3. Let $\phi(z)$ be a continuous function defined along an arbitrary path L (of finite length). Give a direct proof of the statement: the function

$$f(z) = \frac{2!}{2\pi i} \int_L \frac{\phi(\zeta)}{(\zeta - z)^3} \, d\zeta$$

(z being a point not on L) is differentiable and its derivative is given by

$$f'(z) = \frac{3!}{2\pi i} \int_L \frac{\phi(\zeta)}{(\zeta - z)^4} \, d\zeta.$$

*4. Continuation of problem 3. Prove (by mathematical induction) that the ν-th derivative of $f(z)$ ($\nu \geq 1$) is given by

$$f^{(\nu)}(z) = \frac{(\nu + 2)!}{2\pi i} \int_L \frac{\phi(\zeta)}{(\zeta - z)^{\nu + 3}} \, d\zeta.$$

*5. Prove the following somewhat strengthened form of Morera's theorem (KI, p. 66). Let $f(z)$ be a continuous function defined in a simply connected domain $\mathfrak{G}$. Let $\epsilon > 0$ be a given number. Assume that

$$\int_L f(z) \, dz = 0$$

whenever L is the boundary of a rectangle situated in G whose sides are parallel to the coordinate axes and whose diagonals are smaller than ϵ. Then $f(z)$ is regular in $\mathfrak{G}$.

6. Find all values of the integral

$$\int_{0L}^{1} \frac{dz}{1 + z^2}$$

if L may be any path along which the integrand is continuous.

7. Let $\phi(z)$ be a continuous function defined along a closed path C, so that the function

$$f(z) = \frac{1}{2\pi i} \int_C \frac{\phi(\zeta)}{\zeta - z} \, d\zeta$$

is regular within C. It is not necessarily true that $f(z)$ takes on along C the boundary values ϕ. Verify this statement on the example:

$$C = \text{unit circle}, \qquad \phi(z) = \frac{1}{z}.$$

CHAPTER V

EXPANSION IN SERIES

§9. Series with Variable Terms. Uniform Convergence
(KI, 17–19)

1. The general term of the series $\sum_{n=1}^{\infty} f_n(z)$ is given by

*a) $\dfrac{1}{n^z} = e^{-z \log n}$ $(\log n \geq 0)$,

b) $\dfrac{z^n}{1 - z^n}$,

c) $\dfrac{z^n}{1 + z^{2n}}$,

d) $\dfrac{1}{2^n} \dfrac{z^n}{1 - z^n}$,

e) $\dfrac{1}{n^2} \dfrac{z^n}{1 - z^n}$,

f) $\dfrac{a_n z^n}{1 - z^n}$,

g) $\dfrac{\sin nz}{n}$,

h) $\dfrac{\cos nz}{n^2}$,

i) $\dfrac{(-1)^n}{z + n}$,

k) $\dfrac{(-1)^n n}{(z + n) \log n}$

l) $\dfrac{z^{2^n}}{1 - z^{2^{n+1}}}$,

m) $\dfrac{2^n}{z^{2^n} + 1}$.

Find the domain of convergence. (Series a) is a special Dirichlet series, defining the Riemann ζ-function; series b), d), e) are special Lambert series, series f) is the general Lambert series.)

2. Where do the series of problem 1 converge uniformly?

3. If $\Sigma f_n(z)$ converges uniformly and absolutely in

$\mathfrak{G}$ (i.e. if $\Sigma \, |f_n(z)|$ converges uniformly), and if the functions $f_n(z)$ are regular in $\mathfrak{G}$, then $\Sigma f_n'(z)$ converges uniformly and absolutely in every bounded closed subdomain of $\mathfrak{G}$.

4. Let $f_n(z)$ and $h_n(z)$, $n = 0, 1, 2, \cdots$, be regular in $\mathfrak{G}$. Set $s_n(z) = f_0(z) + f_1(z) + \cdots + f_n(z)$. If in $\mathfrak{G}$
a) the limit $\lim_{n\to\infty} s_n(z) \cdot h_{n+1}(z)$ exists uniformly, and
b) the series $\sum_{n=0}^{\infty} s_n(z)[h_n(z) - h_{n+1}(z)]$ converges uniformly, then the series

$$\sum_{n=0}^{\infty} f_n(z) \cdot h_n(z)$$

converges uniformly in $\mathfrak{G}$.

5. Show that the hypotheses of problem 4 are satisfied whenever there exists a number K such that $|s_n(z)| \leq K$ for all z in $\mathfrak{G}$ and all n, and the functions $h_n(z)$ are positive constants monotonically decreasing to $0 : h_n(z) \equiv a_n$, $a_{n-1} \geq a_n \to 0$.

6. Using the theorems stated in the preceding problems, show that the series

$$\sum_{n=1}^{\infty} \frac{(-1)^n}{n^z}$$

converges uniformly in every bounded sub-domain of the half-plane $\Re(z) \geq \delta > 0$.

7. Using the same theorems, show that a *Dirichlet series*, i.e. a series of the form

$$\sum_{n=1}^{\infty} \frac{a_n}{n^z}$$

converges uniformly in every bounded sub-domain of

the half-plane $\Re(z) \geq \Re(z_0) + \delta$, $(\delta > 0)$, provided it converges (or at least possesses uniformly bounded partial sums) for $z = z_0$.

8. Show that the following series possess disjoint domains of convergence, $\mathfrak{G}_1$ and $\mathfrak{G}_2$, in which they represent different functions.

*a) $\displaystyle\sum_{n=1}^{\infty} \frac{1}{n^2} \frac{z^n}{1-z^n}$,

b) $\displaystyle\lim_{n\to\infty} \frac{1}{1+z^n} = \frac{1}{2} + \sum_{n=1}^{\infty} \left(\frac{1}{1+z^n} - \frac{1}{1+z^{n-1}} \right)$,

c)

$$f_2(z) + [f_1(z) - f_2(z)]\left[\frac{1}{2} + \sum_{n=1}^{\infty} \left(\frac{1}{1+z^n} - \frac{1}{1+z^{n-1}} \right) \right]$$

where $f_1(z)$ and $f_2(z)$ are some functions regular for $|z| < 1 + \delta$, $\delta > 0$.

§10. Expansion in Power Series
(KI, 20–21)

1. Compute the first five coefficients of the power series expansion $\Sigma a_n z^n$ of the following functions:

a) $e^{z/(1-z)}$, b) $\sin \dfrac{1}{1-z}$, c) $e^{(e^z)}$,

d) $\log (1 + e^z)$, e) $(\cos z)^{1/2}$, f) $e^{z \sin z}$.

*2. Expand the following functions in a power series $\Sigma a_n z^n$:

a) $\log [a + (a^2 + z^2)^{1/2}]$ where $a > 0$ and $(a^2 + z^2)^{1/2}$ denotes that branch of the function which equals a for $z = 0$.

b) $\dfrac{1}{2} \left(\log \dfrac{1}{1 - z} \right)^2$, c) $(\text{arc tan } z)^2$,

d) $(\text{arc tan } z) \cdot \log (1 + z^2)$, e) $\sin^2 z$ and $\cos^2 z$,

f) $\dfrac{1}{\cos z}$, g) $\dfrac{z}{e^z - 1}$, h) $\log \cos z$, i) $\dfrac{e^z}{e^z + 1}$,

g) $\dfrac{z}{\sin z}$.

3. Expand the following functions in a power series $\Sigma a_n (z - z_0)^n$:

a) $\displaystyle\sum_{n=1}^{\infty} \dfrac{1}{n^z}$, $z_0 = +2$,

b) $\displaystyle\sum_{n=1}^{\infty} \dfrac{z^n}{1 - z^n}$, $z_0 = 0$,

c) $\displaystyle\sum_{n=1}^{\infty} \phi(n) \dfrac{z^n}{1 - z^n}$, $z_0 = 0$.

Here $\phi(1) = 1$, and $\phi(n)$, $n > 1$, denotes the number of positive integers less than and relatively prime to

$n(\phi(2) = 1, \phi(3) = 2, \phi(4) = 2, \phi(5) = 4, \phi(6) = 2,$
$\cdots)$.

4. If $f(z)$ is regular (and not constant) in a simply connected domain $\mathfrak{G}$, then every closed path C in $\mathfrak{G}$ contains in its interior at most a finite number of roots of the equation $f(z) = a$.

5. Does there exist an analytic function $f(z)$ regular at $z = 0$ and assuming at $z = 1, 1/2, 1/3, \cdots$, the following values:

a) $0, 1, 0, 1, \cdots$,

b) $0, \dfrac{1}{2}, 0, \dfrac{1}{4}, 0, \dfrac{1}{6}, \cdots$,

c) $\dfrac{1}{2}, \dfrac{1}{2}, \dfrac{1}{4}, \dfrac{1}{4}, \dfrac{1}{6}, \dfrac{1}{6}, \cdots$,

d) $\dfrac{1}{2}, \dfrac{2}{3}, \dfrac{3}{4}, \dfrac{4}{5}, \cdots$?

6. Let $f(z)$ be regular at z_0 and possess there a zero of order α. How do the functions

$$F_0(z) = \int_{z_0}^{z} f(z)\, dz$$

and

$$F_1(z) = \int_{z_1}^{z} f(z)\, dz$$

behave at this point? (It is assumed that z, z_1 and the path of integration lie in a neighborhood of z_0 in which $f(z)$ is regular and single-valued.)

7. Why is it not correct to say that $z^{1/2}$ has a zero at $z = 0$?

*8. Let $r > 0$ be the radius of convergence of the power series $f(z) = \sum_{n=0}^{\infty} a_n z^n$. Find a lower bound for the absolute values of the zeros of $f(z)$ different from $z = 0$.

§11. Behaviour of Power Series on the Circle of Convergence

(KI, 24)

(Problems 5–9, 14 and 15 from §4 dealt with similar questions.)

1. Give examples of power series converging
a) at all points on the circle of convergence,
b) at no such point,
c) at all points except one.

2. At which points of their circles of convergence do the following power series converge?

a) $\displaystyle\sum_{n=2}^{\infty} (-1)^n \frac{z^n}{n \log n}$,   b) $\displaystyle\sum_{n=2}^{\infty} (-1)^{n-1} \frac{z^{2n}}{n - n^{1/2}}$,

c) $\displaystyle\sum_{n=0}^{\infty} (-1)^n \frac{z^{2n+1}}{2n + 1}$,   d) $\displaystyle\sum_{n=3}^{\infty} \frac{z^n}{\log \log n}$.

*3. Let all coefficients of the series $\sum_{n=0}^{\infty} a_n z^n$ be real and non-negative, and let $r > 0$ be its radius of convergence. Show that $z_1 = +r$ is a singular point of the function $f(z) = \Sigma a_n z^n$. (Hint: expand $f(z)$ about $z = r/2$.)

*4. Show that the following series cannot be continued beyond their circles of convergence:

a) $\displaystyle\sum_{n=0}^{\infty} z^{n!}$,   b) $\displaystyle\sum_{n=0}^{\infty} z^{(2^n)}$,   c) $\displaystyle\sum_{n=0}^{\infty} z^{(g_1 g_2 \cdots g_n)}$

where g_1, g_2, $\cdots$ are arbitrary positive integers ≥ 2,

d) $\displaystyle\sum_{n=0}^{\infty} \frac{z^{2^n+2}}{(2^n + 2)(2^n + 1)}$.

(Hint: use the results of the preceding problem or the proof given in KI, p. 101.)

*5. The functions

a) $f(z) = \displaystyle\sum_{n=1}^{\infty} \frac{z^n}{1 - z^n}$,   b) $g(z) = \displaystyle\sum_{n=1}^{\infty} \frac{z^n}{1 + z^{2n}}$

are defined and regular for $|z| < 1$. Show that they can not be continued beyond the unit circle.

*6. The power series $\sum_{n=1}^{\infty} \{(-1)^{[n^{1/2}]}/n\} z^n$ converges (conditionally) at all points of its circle of convergence $|z| = 1$. $[n^{1/2}]$ denotes the largest integer not exceeding $n^{1/2}$. (Hint: Consider first the point $z = 1$; then use the result of §3, problem 14.)

7. If $r = 1$ is the radius of convergence of the series $h(z) = \Sigma b_n z^n$, if $b_n > 0$ and Σb_n diverges, then

$$\lim_{x \to +1} h(x) = +\infty$$

where x denotes a real variable.

8. Let Σa_n be a convergent series. Using the result of §9, problem 4, show that the power series $\Sigma a_n z^n$ converges uniformly on the real segment $0 \leq x \leq 1$.

*9. Using the result of §1, problem 13, show that the power series of problem 8 converges uniformly within the triangle z_1, z_2, 1, provided the vertices z_1 and z_2 lie within the unit circle.

*10. Prove Abel's theorem on power series: If Σa_n converges, then

$$\lim_{z \to +1} \sum a_n z^n = \sum a_n$$

provided z approaches 1 within a triangle z_1, z_2, 1 with the vertices z_1 and z_2 situated within the unit circle.

CHAPTER VI

CONFORMAL MAPPING

§12. Linear Functions. Stereographic Projection

1. The transformations

a) $w = 3z + 5i$, b) $w = \dfrac{i}{2}(z + 3)$, c) $w = az + b$,

are rotations about a point (fixed point of the transformation) followed by a magnification. Find the fixed points, rotations and magnifications (i.e. write the transformations in the form $w - z_0 = a_0(z - z_0)$).

2. Find an entire linear transformation which takes the triangle $\Delta_z = (0, 1, i)$ in the z plane into the triangle $\Delta_w = (-1, -i, +i)$ in the w plane. Is the transformation uniquely determined?

3. Show that every similarity transformation is given by an entire linear function.

4. Find the images of the following figures under the transformation $w = 1/\bar{z}$ (reflection with respect to the unit circle).

a) The circle $|z - 1| = 1$,

b) The circle $|z - 1/2| = 1/4$,

c) The circle $|z| = r$

d) The circle $|z - z_0| = |z_0| > 0$,

e) The circle $|z - z_0|^2 = |z_0|^2 - 1 > 0$,

f) The circle $\alpha(x^2 + y^2) + \beta x + \gamma y + \alpha' = 0$,

g) The straight line $\Re(z) = \alpha (\gtreqless 0)$,

h) The family of all straight lines parallel to the bisector of the first quadrant,

i) The family of all straight lines parallel to the segment $0 \cdots z_0 (z_0 \neq 0)$,

k) The family of all straight lines passing through $z_0 \neq 0$,

l) The family of all circles passing through z_0 ,

m) The family of all circles passing through z_0 and $z_1 \neq z_0$,

n) The triangle z_1 , z_2 ; z_3 ,

o) The parabola $y^2 - 2px = 0$,

p) The hyperbola $x^2 - y^2 - 1 = 0$,

q) The parabola $y - x^2 = 0$.

5. Answer the questions of the preceding problem for the case of a reflection with respect to an arbitrary circle $| z - a | = \rho$. (The reflection of a point z is the point z' such that am $(z' - a) = $ am $(z - a)$, $| z' - a | \cdot | z - a | = \rho^2$. We have $(z' - a)/\rho = \rho/(\bar{z} - \bar{a})$.)

Preliminary remarks to problems 6-10.

These problems deal with the stereographic projection of a sphere of radius 1 on the plane. We call the point of contact between the sphere and the plane (i.e. $z = 0$) South Pole, the diametrically opposite point (image of $z = \infty$) North Pole and use the geographical termini like meridian, equator, etc. As prime meridian we take the image of the positive real axis. A point of the sphere may be identified by its latitude and longitude.

6. Find the images on the sphere of the following points, lines and regions in the plane.

a) $+1, +i, -1, -i, z = x + iy$,

b) $| z | < 1, | z | = 1, | z | > 1$,

c) $\Re(z) > 0, = 0, < 0$,

d) $\Im(z) > 0$, $=0$, <0,

e) $|z| = $ const.

f) am $z = $ const.

7. Find the relative position of the images on the sphere of

a) z and $-z$

b) z and $\bar{z}$

c) z and $1/\bar{z}$

d) z and $1/z$.

8. Consider in the plane:

a) a pencil of parallel straight lines,

b) a reflection with respect to the x-axis,

c) a reflection with respect to the y-axis,

d) a reflection with respect to the unit circle,

e) a triangle.

What figures or transformations on the sphere are obtained from these by the stereographic projection?

9. Find the images in the plane of the following figures on the sphere:

a) meridian of longitude λ, latitude circle of latitude β,

b) two andipodal points,

c) great circles,

d) a spherical triangle,

e) the spherical center M_0 of a circle k,

f) the pencil of great circles passing through a point P,

g) the point of longitude λ and latitude β.

*10. Consider the projection from the center of the sphere on the plane. This yields a one-to-one mapping of the southern hemisphere on the plane. a) Find the analytic expression for the mapping of the plane onto the hemisphere. b) Where is this mapping angle-preserving? c) Where is this mapping length-preserving?

11. Find the general form of a linear transformation with two distinct fixed points ζ_1 and ζ_2 .

12. Find the linear transformation $w = (az + b)/(cz + d)$ which takes three given distinct points z_1 , z_2 , z_3 into three given distinct points w_1 , w_2 , w_3 .

13. Let there be given a circle k_z , a point z_1 and its reflection with respect to k_z , the point z_2 . The linear transformation $w = (az + b)/(cz + d)$, $ad - bc \neq 0$ takes k_z into a circle k_w , z_1 into w_1 , z_2 into w_2 . Show that w_2 is the reflection of w_1 with respect to k_w .

14. Find a linear transformation which maps the domain interior to the unit circle onto the domain $|w - 1| < 1$ and takes $z = 0$ and $z = 1$ into $w = 1/2$ and $w = 0$, respectively. Is this transformation uniquely determined?

*15. Find the general form of a linear transformation which maps the upper half-plane onto itself.

16. The domain $|z| < 1$ is mapped onto the upper half-plane by a linear transformation which takes 1, i, -1 into 0, 1, ∞, respectively. Find the mapping. What are the images of the radii of the unit circle, in particular of the radii leading to the points 1, i, -1, $-i$?

17. The linear transformation taking 1, i, -1 into i, 0, $-i$ respectively maps the domain exterior to the unit circle onto the right half-plane. Find the mapping. What are the images of the rays am $z = $ const., $|z| \geq 1$? What are the images of the circles $|z| = r > 1$?

*18. Find the general form of a linear transformation mapping the domain interior to the unit circle onto itself.

*19. Find the form of a linear transformation which

corresponds to the rotation of the sphere by means of a stereographic projection.

20. Given two non-concentric circles $k_1 : |z - z_1| = r_1$, $k_2 : |z - z_2| = r_2$ possessing no points in common, find a linear transformation taking k_1 and k_2 into two concentric circles k_1' and k_2' with center at $w = 0$. Is this transformation uniquely determined? Which point in the z-plane is taken into $w = 0$?

21. (Continuation of problem 11.) a) Find the general form of a linear transformation $w = (az + b)/(cz + d)$ possessing a single fixed point ζ_1.
b) Using the results of problem 11 and 21a investigate the behaviour, under a linear transformation, of circles through the fixed points and their orthogonal circles.

*22. Given the linear transformation $w = (az + b)/(cz + d)$, $ad - bc \neq 0$, and a point z_0, set

$$z_{\nu+1} = \frac{az_\nu + b}{cz_\nu + d}, \qquad \nu = 0, 1, \cdots .$$

Investigate the sequence (z_ν). Does it converge? Does it contain infinitely many points?

§13. Simple Non-Linear Mapping Problems

1. Discuss the mapping of the period-strip $-\pi < \Im(z) \leq +\pi$ by the function $w = e^z$. Find the images of the straight segments $\Re(z) = $ const., and of the straight lines $\Im(z) = $ const.

2. Discuss the mapping of the period-strip $-\pi < \Re(z) \leq +\pi$ by the function $w = \sin z$. Find the images of the straight lines $\Re(z) = $ const. and of the straight segments $\Im(z) = $ const.

3. Discuss in the same manner the mapping of the period-strip $-\pi/2 < \Re(z) \leq +\pi/2$ by the function $w = \tan z$.

4. Find the images of the segments and lines mentioned in problem 1 under the transformation

$$w = \operatorname{sh} z = \frac{e^z - e^{-z}}{2}.$$

5. Find the image of the doubly connected region $1 < |z| < +\infty$ under the transformation $w = z + 1/z$.

6. Find the images of the doubly connected regions $0 < |z| < 1$ and $0 < |z| < +\infty$ under the transformation $w = z + 1/z$.

7. Find the images of the intersections of the closed upper half-plane $\Im(z) \geq 0$ and the closed regions

a) $1 \leq |z| \leq 2$,   b) $\frac{1}{2} \leq |z| \leq 1$,   c) $\frac{1}{2} \leq |z| \leq 2$

under the transformation $w = z + 1/z$.

8. Find the images of
a) the sector $|z| \geq 1$, $0 \leq \operatorname{am} z \leq \pi/3$,
b) the sector $0 \leq \operatorname{am} z \leq \pi/3$,
c) the sector $|z| \geq 1$, $-\pi/3 \leq \operatorname{am} z \leq +\pi/3$,
under the transformation $w = z + 1/z$.

9. Map the domain exterior to the ellipse $|z - 2| + |z + 2| = 100/7$ onto the interior of the unit circle. (Hint: Use the properties of the function $w = z + 1/z$.)

10. Map the sector $|z| \leq 1$, $0 \leq \operatorname{am} z \leq \pi/3$ onto the unit disc.

11. Map the following domains onto the unit disc:

a) The intersection of $|z| \leq 1$ and $|z - 1| \leq 1$.

b) The closed region $|z| \leq 1$, $\Im(z) \geq 0$.

*12. Map the intersections of the closed regions $|z| \leq 1$ and $|z - 1/2| \geq 1/2$ onto the unit disc $|w| \leq 1$.

*13. Map the half-strip $-1/2 \leq \Re(z) \leq 1/2$, $\Im(z) \geq 0$ onto the unit disc.

*14. Map the domain bounded by the parabola $\Im(z^{1/2}) = \alpha > 0$ (and containing the origin) onto the interior of the unit circle.

*15. Show that the principal value of the logarithm satisfies the following inequalities:

a) $\quad |\log z| \underset{>}{\lessgtr} |\log (1 - z)|$ when $|z| \underset{<}{\gtrless} |1 - z|$,

$\quad$ i.e. when $\Re(z) \underset{<}{\gtrless} \dfrac{1}{2}$,

b) $\quad |\log z| \underset{>}{\lessgtr} \left| \log \dfrac{z - 1}{z} \right|$ when $|z - 1| \underset{>}{\lessgtr} 1$,

c) $\quad |\log (1 - z)| \underset{>}{\lessgtr} \left| \log \dfrac{z - 1}{z} \right|$ when $|z| \underset{>}{\lessgtr} 1$.

*16. Using the results of the preceding problem, determine in which parts of the plane one of the three numbers $|\log z|$, $|\log (1 - z)|$, $|\log (z - 1)/z|$ is smaller than the other two. (The symbol log denotes here the principal value of the logarithm.)

Part II—ANSWERS

FUNDAMENTAL CONCEPTS

§1. Numbers and Points

1. a) $-z_0$, b) $\bar{z}_0$, c) $-\bar{z}_0$, d) $i\bar{z}_0$, e) $-i\bar{z}_0$.

2. The proof follows by squaring both sides of both inequalities.

3.e) and f) Hyperbolas (pairs of stright lines for $\alpha = 0$). To verify the statement, set $z = x + iy$. g) and h) Lemniscates; in fact $|z^2 - z| = |z| \cdot |z - 1|$ is the product of the distances of z from 0 and 1, $|z^2 - 1|$ is the product of the distances of z from 1 and -1. k) The right half-plane including its boundary. l) and m) Appollonius' circles. n) The perpendicular bisector of the segment $z_1 \cdots z_2$.

4. If and only if the difference quotient is real, since the direction of the numerator (from z_3 to z_1) must be equal or opposite to that of the denominator (from z_3 to z_2).

5. If and only if the cross-ration is real. In fact, the four points are collinear if and only if both $(z_1 - z_3)/(z_2 - z_3)$ and $(z_1 - z_4)/(z_2 - z_4)$ are real. If the four points are not collinear, the amplitude of $(z_1 - z_3)/(z_2 - z_3)$ can differ from that of $(z_1 - z_4)/(z_2 - z_4)$ only by a multiple of π.

6. The identity is proved by setting $z = x + iy$. It

41

expresses the fact that in a parallelogram the sum of the squares of the diagonals equals the double product of two adjacent sides.

7. $$z = \frac{\lambda_2 z_1 + \lambda_1 z_2}{\lambda_1 + \lambda_2}.$$

8. a) $\dfrac{z_1 + z_2 + z_3}{3}$,　　b) $\dfrac{\lambda_1 z_1 + \lambda_2 z_2 + \lambda_3 z_3}{\lambda_1 + \lambda_2 + \lambda_3}.$

c) Show that in 7 z lies between z_1 and z_2 whenever $\lambda_1 > 0$, $\lambda_2 > 0$, and apply this result twice.

9. The proof follows by mathematical induction from the results of problems 7 and 8.

10. Consider the polynomial $(z - z_1) \cdot (z - z_2) \cdot (z - z_3) \equiv z^3 + a_1 z^2 + a_2 z + a_3$. We have $a_1 \equiv -(z_1 + z_2 + z_3) = 0$. Since $|z_j| = 1$, $1/z_j = \bar{z}_j$, $j = 1, 2, 3$. Thus

$$a_2 = z_1 z_2 + z_1 z_3 + z_2 z_3 = z_1 z_2 z_3 (\bar{z}_1 + \bar{z}_2 + \bar{z}_3) = 0.$$

It follows that the z_j are roots of an equation $z^3 - a_3 = 0$, $|a_3| = 1$. Hence $z_2 = c z_1$, $z_2 = c^2 z_1$ where $c = (\cos 120° + i \sin 120°)$.

11. Using the same method, we can show that the z_j $(j = 1, 2, 3, 4)$ are roots of an equation $z^4 + a_2 z^2 + a_4$. Hence there exist two numbers a and b, such that $|a| = |b| = 1$, $z_1 = -z_3 = a$, $z_2 = -z_4 = b$.

12. If and only if $(z_1 - z_3)/(z_2 - z_3) = (z_1' - z_3')/(z_2' - z_3')$. In fact, this equation expresses the fact that the angles at z_3 and z_3' are equal, and the sides $z_1 z_3$, $z_2 z_3$ and $z_1' z_3'$, $z_2' z_3'$ are proportional.

13. a) Choose a number r such that $r < 1$, $r \geq |z_1|$, $r \geq |z_2|$. For $|z| \leq r$ we have $(|1 - z|)/$

$(1 - |z|) \leq (1 + r)/(1 - r) = K_1$. Now we consider the triangle $\Delta = (1, z_3, \bar{z}_3)$, z_3 being the point on $|z| = r$ such that the segment $z_3 \cdots 1$ is tangent to this circle. Points z of the triangle $1, z_1, z_2$ with $|z| \geq r$ lie within the (closed) triangle Δ. Set $r_1 = (1 - r^2)^{1/2}$. Then $r_1 = |1 - z_3|$. We denote the angle in Δ at 1 by $2\phi_0$ and note that $0 < \phi_0 < \pi/2$, $r_1 = \cos \phi_0$. All points of Δ (except $z = 1$) are of the form $z = 1 - \rho (\cos \phi + i \sin \phi)$ with $0 < \rho \leq \cos \phi_0$, $|\phi| \leq \phi_0$. We have $-2\rho \cos \phi + \rho^2 \leq -\rho \cos \phi_0 + 1/4\rho^2 \cos^2 \phi_0$, so that

$$\frac{\rho}{1 - (1 - 2\rho \cos \phi + \rho^2)^{1/2}} \leq \frac{2}{\cos \phi_0},$$

i.e.

$$\frac{|1 - z|}{1 - |z|} \leq \frac{2}{\cos \phi_0} = K_2.$$

To complete the proof we must merely choose for K the larger of the two numbers K_1 and K_2.

b)
$$K = \frac{1}{2^{1/2} - 1}.$$

§2. Point Sets. Paths. Regions

1. All equations of the type considered may be arranged in a sequence, since for every positive integer k there is only a finite number of equations such that $k \leq n + |a_n| + |a_1| + \cdots + |a_n| < k + 1$. Since every equation has at most n distinct roots, the assertion follows.

2. The set of all real rational numbers is countable. In fact the array

$$\frac{1}{1}, \frac{2}{1}, \frac{3}{1}, \cdots, \frac{n}{1}, \cdots$$

$$\frac{1}{2}, \frac{2}{2}, \frac{3}{2}, \cdots, \frac{n}{2}, \cdots$$

$$\frac{1}{3}, \frac{2}{3}, \frac{3}{3}, \cdots, \frac{n}{3}, \cdots$$

$$\cdots\cdots$$

contains all real rational numbers. This array may be rearranged into the sequence

$$1, 2, \frac{1}{2}, 3, \frac{1}{3}, 4, \frac{3}{2}, \frac{2}{3}, \frac{1}{4}, 5, \cdots$$

using the so-called "diagonal method." (We write down first all numbers p/q with $p + q = 2$, then all numbers with $p + q = 3$, etc., omitting any number already appearing in the sequence.)

Now let r_1, r_2, $\cdots$ be a sequence of all real rational numbers. The array

$$r_1 + ir_1, r_1 + ir_2, \cdots, r_1 + ir_n, \cdots$$

$$r_2 + ir_1, r_2 + ir_2, \cdots, r_2 + ir_n, \cdots$$

$$r_3 + ir_1, \cdots$$

$$\cdots\cdots\cdots .$$

contains all elements of the set considered. It may be rearranged into a sequence by the diagonal method.

3. The proof follows closely the one given above.

4. The four numbers α, λ, μ, β are given below. Numbers belonging to the set considered are marked by asterisks.

α) $-10^{1/2}$; $-10^{1/2}$; $+10^{1/2}$; $+10^{1/2}$;

β) 0^*, $\dfrac{1}{e}$, e, e;

c) 0^*, 1, 1, 2^*;

d) 0^*, $+\infty$, $+\infty$, $+\infty$;

e) $\dfrac{2^*}{3}$, $+\infty$, $+\infty$, $+\infty$;

f) 0, 0, 1^*, 2^*;

g) 0, 0, e, 4^*;

h) -2^*, -1^*, $+1^*$, $+2^*$;

i) -1^*, 0, 2, $\dfrac{5^*}{2}$;

k) $\dfrac{1^*}{9}$, $\dfrac{1^*}{9}$, 1^*, 1^*.

5. Our set may be constructed in the following way. Divide the interval $0 \cdots 1$ into 10 intervals of equal

length and erase the interior points of the 1st, 3rd, 5th, 7th, and 9th subinterval, as well as the point 0. Divide each of the remaining 5 intervals of length 1/10 into ten intervals of equal length and erase all interior points of the 1st, 3rd, 7th, and 9th subintervals of length 1/100, as well as the left end points of the 5 intervals of length 1/10. Now repeat this procedure with the remaining 25 intervals of length 1/100 and continue this procedure *ad infinitum*. The remaining points form our set. In fact, after p steps those and only those points are left which admit the infinite decimal expansion $.\alpha_1\alpha_2 \cdots \alpha_p\alpha_{p+1} \cdots$ with α_1, $\cdots \alpha_p$ odd.

Now let x_0 be a point of the set. After p steps x_0 belonged to a non-erased interval of length $1/10^p$. It follows that every neighborhood of x_0 contains other points of the set. This proves the assertion.

The set constructed in this problem (a so-called *Cantor* set) possesses many interesting properties. Observe, for instance, that at each step we erase half of the intervals, so that the total lengths of the 5^n intervals remaining after n steps is $1/2^n$. The total length approaches zero as $n \rightarrow \infty$. Nevertheless the set contains all numbers. $\alpha_1\alpha_2 \cdots$ with odd α_i. (Note that a number like .352 belongs to the set since $.352 = .3519999 \cdots$.)

6. If α does not belong to the set, then each interval $\alpha \cdots \alpha + \epsilon$ $(\epsilon > 0)$ contains at least one point of the set, hence infinitely many such points. α is a limit point, so that $\alpha = \lambda$. The same argument holds for β and μ.

7. No, since the set contains all real negative num-

bers. It covers the region interior to the parabola $y^2 = 1 - 2x$ and the parabola itself.

8. a) $z = 0, +1/m, + i/n$;
   b) all z for which $|z| \leq 1$;
   c) all points of the plane;
   d) all points of the plane;
   e) all z for which $|z| \leq 1$.

9. Yes. This follows from the construction of the set given above (see problem 5). Whenever a point x_0 is being erased, it is either an interior point of an interval which is being erased, or an isolated left end point. In the latter case either $x_0 = 0$, or the points immediately to the left of x_0 have been erased previously. It follows that a sufficiently small neighborhood of a point x_0 not in the set contains no points of the set. Only points of the set can be limit points.

10. Yes. (See definition in KI, p. 7)

11. If ζ belongs to M and a circle about ζ encloses no points of M', it encloses only points of M, so that ζ is an interior point of M.

12. Let R be the set of boundary points of M, Z a limit point of R. For every $\rho > 0$ the neighborhood $|z - Z| < \rho/2$ contains at least one point z_0 of M. This point z_0 lies in the neighborhood $|z - Z| < \rho$. Thus Z is a limit point of M. As a limit point of R it can not be an interior point of M. It follows that Z belongs to R.

13. Assume there is no number d possessing such a property. Then we can find a sequence (z_n') of points of M' and a sequence (z_n'') of points of M'' such that $|z_n' - z_n''| \leq 1/n$. Since M' is bounded, so is (z_n')

Thus there exists at least one limit point ζ of (z'_n). ζ also is a limit point of (z''_n). M' and M'' being closed, ζ belongs to both M' and M''. This contradicts our hypothesis.

Now let d_0 be the least upper bound of all such d. d_0 is finite and possesses the required property. Thus d_0 is the largest d. (d_0 is the *distance* between M' and M''; cf. KI, p. 8.)

14. Let E_k denote the point of the curve for which $|y| = x = 2/(2k - 1)$, N_k the point for which $y = 0$, $x = 1/k$. The length of the polygonal path $N_{k-1}E_kN_k$ exceeds the double ordinate of E_k ; hence it exceeds $2/k$. Since the series $\Sigma\, 1/k$ diverges, it follows that in an arc of the curve containing the origin we can inscribe a "segmental arc" (see KI, p. 17.) of arbitrarily large length.

15. $\mathfrak{M}$ is a domain. In fact, let $z_0 = x_0 + iy_0$ be a point of $\mathfrak{M}$. If $y_0 > 1$, then all points z with $|z_0 - z| < y_0 - 1$ are points of $\mathfrak{M}$. If $|y_0| \leq 1$, there exists a positive number δ such that $|x_0| > \delta$ and $|x_0 \pm 1/n| > \delta$, $n = 1, 2, \cdots$. If $0 < \rho < \delta$ and $\rho < y_0$, then all points z with $|z_0 - z| < \rho$ are points of $\mathfrak{M}$. Now let $z' = x' + iy'$ and $z'' = x'' + iy''$ be two points of $\mathfrak{M}$. Let $p > 1$ be a real number such that $p > y'$, $p > y''$. z' and z'' may be joined by the path (lying in $\mathfrak{M}$) which consists of the straight segments $z' \cdots x' + ip, x' + ip \cdots x'' + ip, x'' + ip \cdots z''$.

All points on the real axis and all points on the excluded segments are boundary points of $\mathfrak{M}$. There are no other boundary points.

$i/2$ is an *inaccessible* boundary point. There exists no path (in $\mathfrak{M}$) leading from z_0 to $i/2$. In fact every such path $\mathfrak{p}$ would have to contain infinitely many pairs of points $z' = x' + iy'$, $z'' = x'' + iy''$ such that $|z' - i/2| < 1/4$, $|z'' - i/2| < 1/4$ and x', x'' are not both contained in an interval $(1/n, 1/(n + 1))$. The length of any path in $\mathfrak{M}$ joining z' and z'' must exceed $1/2$. It follows that $\mathfrak{p}$ can not be of finite length.

16. Yes. In fact, S is a Jordan arc (one-to-one continuous image of the segment $1 \geq t \geq 0$), and an elementary computation shows that the length of S is $2^{1/2}/e$.

17. Consider first a bounded plane region $\mathfrak{G}$. Let M be its boundary set. If M is not connected, there exists a simple closed path $\mathfrak{p}$ containing points of M in both its interior and exterior regions. If such a path exists, M is not connected, since the parts of M within and outside of $\mathfrak{p}$ are closed. This shows that our assertion is correct for bounded plane regions. The same argument holds on the sphere where all regions are bounded.

18. Yes. The boundary of $\mathfrak{M}$ is connected.

19. Let $\mathfrak{p}$ be a simple closed path separating the two points not in $\mathfrak{M}$. Since $\mathfrak{M}$ is simply connected, $\mathfrak{p}$ contains points not belonging to $\mathfrak{M}$. Since this is true for any such $\mathfrak{p}$, the assertion follows.

CHAPTER II

INFINITE SEQUENCES AND SERIES

§3. Limits of Sequences. Infinite Series with Constant Terms

1. The definition of a limit point implies that the domain $0 < |z - \zeta| < 1$ contains a point z_n, say z_{n_1}, that the domain $0 < |z - \zeta| < 1/2$ contains a point z_n with $n > n_1$, say z_{n_2}, and in general that for $\nu \geq 2$, the domain $0 < |z - \zeta| < 1/\nu$ contains a point z_n with $n > n_{\nu-1}$, say z_{n_ν}. The sequence of numbers $z'_\nu = z_{n_\nu}$ converges to ζ.

2. The theorem and proof are obtained from those of problem 3 by setting $p_n = 1$. The theorem does not hold for $\zeta = \infty$. Example: set $z_{2k-1} = -z_{2k} = 2k - 1$, $k = 1, 2, \cdots$. The sequence $z_1, z_2, \cdots$, i.e. the sequence $1, -1, 3, -3, 5, -5, \cdots$ converges to ∞; the sequence $z'_1, z'_2, \cdots$, i.e. the sequence $1, 0, 1, 0, 1, 0, \cdots$ has no limit. The theorem is true in the real domain, provided a definite sign is attached to the symbol ∞ : if (z_n) is a sequence of real numbers and $z_n \to +\infty$ (or $z_n \to -\infty$), then $z'_n \to +\infty$ (or $z'_n \to -\infty$).

3. If $\zeta = 0$, then for every $\epsilon > 0$ we can choose an m such that $|z_\nu| < \epsilon/2$ for $\nu > m$, and a $n_0 > m$ such that $|p_1 z_1 + \cdots + p_m z_m|/P_n < \epsilon/2$, for $n > n_0$. For $n > n_0$ we have $|z'_n| < \epsilon$. If $\zeta \neq 0$, apply the preceding reasoning to the sequence $(z_n - \zeta)$.

4. The proof is similar to the preceding one. Assume that $\zeta = 0$ and that ϵ is a given positive number. Choose an m such that $|z_\nu| < \epsilon\beta/2$ for $\nu > m$, and

a $n_0 > m$ such that $|b_1 z_1 + \cdots + b_m z_m|/|b_1 + \cdots + b_n| < \epsilon/2$ for $n > n_0$. This is possible, for the numerator of the preceding fraction is constant and the denominator converges to infinity since $|b_1 + \cdots + b_n| \geq \beta(|b_1| + \cdots + |b_n|)$. For $n > n_0$ we have that

$$|z_n'| \leq \frac{\epsilon}{2} + \frac{1}{2}\epsilon\beta \frac{|b_{m+1}| + \cdots + |b_n|}{|b_1 + \cdots + b_n|} \leq \epsilon.$$

If $\zeta \neq 0$, apply the preceding reasoning to the sequence $(z_n - \zeta)$.

5. Given a positive ϵ, choose an m such that $|z_\nu| < \epsilon/2M$ for $\nu > m$. If $n > m$, condition 2 implies that

$$|z_n'| \leq |a_{n1}| \cdot |z_1| + |a_{n2}| \cdot |z_2| + \cdots$$
$$+ |a_{nm}| \cdot |z_m| + \frac{\epsilon}{2}.$$

If m is kept fixed and $n \to \infty$, each term in the finite sum converges to 0 (condition 1). It follows that there exists a $n_0 > m$ such that this sum is less than $\epsilon/2$ for $n > n_0$. Hence $|z_n'| < \epsilon$ if $n > n_0$.

6. We have

$$z_n' - A_n\zeta = a_{n1}(z_1 - \zeta) + a_{n2}(z_2 - \zeta) + \cdots$$
$$+ a_{nn}(z_n - \zeta).$$

Since $(z_n - \zeta) \to 0$, the right hand side $\to 0$ (problem 5). Since $A_n\zeta \to \zeta$ (condition 3), it follows that $z_n' \to \zeta$. The theorem of problem 4 is obtained by setting $a_{np} = b_p/(b_1 + b_2 + \cdots + b_n)$, $n = 1, 2, \cdots, 1 \leq p \leq n$, that of problem 3 by assuming the b_n to be positive real numbers, that of problem 2 by setting all b_n equal to 1.

7. a) The hypothesis implies that $z_n' z_n'' \to \zeta' \zeta''$; the assertion follows from the result of problem 2.

b) We have

$$z_n = \frac{z_n''(z_1' - \zeta') + z_{n-1}''(z_2' - \zeta') + \cdots + z_1''(z_n' - \zeta')}{n}$$

$$+ \zeta' \frac{z_1'' + z_2'' + \cdots + z_n''}{n}.$$

The second term on the right hand side $\to \zeta' \zeta''$ (problem 2). The first term $\to 0$. This follows from problem 5 by setting $a_{np} = z_{n-p+1}''/n$. These numbers satisfy conditions 1) and 2), for the sequence (z_n'') is bounded.

c) The proof is similar to that of b). We have $z_n = [a_{n1}z_n''(z_1' - \zeta') + a_{n2}z_{n-1}''(z_2' - \zeta') + \cdots + a_{nn}z_1''(z_n' - \zeta')] + \zeta'[a_{nn}z_1' + a_{n,n-1}z_2' + \cdots + a_{n1}z_n'']$. The numbers $a_{np}z_{n-p+1}''$ satisfy the conditions imposed in problem 5 on the numbers a_{np}, since the sequence (z_n'') is bounded. Hence the first term $\to 0$. The numbers $a_{n,n-p+1}$ satisfy the conditions imposed on the numbers a_{np} in problems 5 and 6 (condition 4). It follows that the second term $\to \zeta' \zeta''$.

8. The assertion follows from that of problem 6 by setting $a_{np} = \frac{1}{2^n}\binom{n}{p}$. Condition 2) is satisfied since $\binom{n}{p} < n^p$, condition 2) is verified by setting $M = 1$ and condition 3) holds since $\binom{n}{1} + \binom{n}{2} + \cdots + \binom{n}{n} = 2^n - 1$.

9. Set $a_{np} = z_{n-p+1}''$ and apply the result of problem 5.

10. a) Assume that the series converges. For every

positive ϵ there exists a n_0 such that for $n > n_0$ and all $p > 0$

$$| c_{n+1} + c_{n+2} + \cdots + c_{n+p} | < \epsilon.$$

If $n > n_0$, then $| T_n | < \epsilon$. Hence $T_n \to 0$, and the condition is seen to be necessary.

b) Assume that the series diverges. Then there exists a number $\epsilon_0 > 0$ such that for any number n_0 there exists at least one (and hence infinitely many) pairs of numbers $n > n_0$ and p such that

$$| c_{n+1} + c_{n+2} + \cdots + c_{n+p} | \geq \epsilon_0 .$$

Therefore there exists a sequence of positive numbers p_n such that the corresponding sequence T_n does not converge to zero. This shows that the condition is sufficient.

11. To prove this identity set $a_\nu = s_\nu - s_{\nu-1}$ (for $\nu = 0, a_0 = s_0$) and collect terms with the same s_ν.

12. By virtue of problem 12

$$T_n = \sum_{\nu=n+1}^{n+p_n} a_\nu b_\nu = \sum_{\nu=n+1}^{n+p_n} s_\nu (b_\nu - b_{\nu+1})$$

$$- [s_n b_{n+1} - s_{n+p_n} b_{n+p_n+1}].$$

The hypotheses and problem 10 shows that $T_n \to 0$ for every choice of the positive numbers p_n. This implies (problem 10) that $\Sigma a_n b_n$ converges.

13. a) Since the s_n are bounded and $b_n \to 0$, $s_n b_{n+1} \to 0$. Since $\sum_{\nu=0}^{n} (b_\nu - b_{\nu+1}) = b_0 - b_{n+1}$, the series $\Sigma (b_n - b_{n+1})$ converges; it converges absolutely since $b_n > b_{n+1}$. The sequence s_n being bounded, it follows that $\Sigma s_n (b_n - b_{n+1})$ converges.

b) $a_{n-1} > b_n > \alpha$ implies that $b_n \to \alpha_0 \geq \alpha$. Since the sequence (s_n) converges, so does the sequence $s_n b_{n+1}$. Since $\sum_{\nu=0}^{n} (b_\nu - b_{\nu+1}) = b_0 - b_n \to b_0 - \alpha_0$, the series $\Sigma(b_n - b_{n+1})$ converges absolutely and so does $\Sigma s_n(b_n - b_{n+1})$ (for the sequence (s_n) is bounded).

c) $s_n b_{n+1} \to 0$. The convergence of $\Sigma s_n(b_n - b_{n+1})$ follows by the argument used in the first two cases.

14. As in problem 11 we have

$$\sum_{\nu=n+1}^{n+p} a_\nu b_\nu = \sum_{\nu=n+1}^{n+p} s_\nu(b_\nu - b_{\nu+1}) - s_n b_{n+1} + s_{n+p} b_{n+p+1},$$

$$\left| \sum_{\nu=n+1}^{n+p} a_\nu b_\nu \right| \leq K \left\{ \sum_{\nu=n+1}^{n+p} \nu^{1/2}(b_\nu - b_{\nu+1}) + n^{1/2} b_{n+1} \right.$$

$$\left. + (n+p)^{1/2} b_{n+p+1} \right\}.$$

It follows from the hypotheses that $\left| \sum_{\nu=n+1}^{n+p} a_\nu b_\nu \right|$ will be less than any preassigned positive number ϵ if n is sufficiently large.

§4. Convergence Properties of Power Series

1. a) For all four series $r = 1$;

b) $r = +\infty$, $r = 1$, $r = 0$;

c) $r = c$ (since $a_n^{1/n} = n^{-1} n!^{1/n} \to 1/e$),

$r = 0$ (since $a_n^{1/n} = \sim e^{-2} n \to +\infty$),

$r = e$ (since $a_n^{1/n} = (1 - 1/n)^n \to 1/e$),

$r = 1$ (since $a_n^{1/n} = e^{(\log n)^2/n} \to e^0 = 1$);

d) $r = 1$ (since $2 \leq \tau(n) \leq n$ for $n \geq 2$),

$r = 1$ (since $1 \leq \phi(n) \leq n$);

e) $r = 1/3^{1/2}$ (since $\overline{\lim} \, a_n^{1/n} = \lim (3^k)^{1/(2k+1)} = 3^{1/2}$),
$r = 1$ (since $1 \leq a_n \leq n^2$), $r = 0$ (since

$$\overline{\lim} \, a_n^{1/n} = \lim (\log k)^{k/(2k+1)}.$$

2. a) If $|z| < |z_0|$, then $|z/z_0| = \theta < 1$, $|a_n z^n| = |a_n z_0^n| \cdot \theta^n < K \cdot n^k \cdot \theta^n$. The series $\Sigma n^k \cdot \theta^n$ converges for $|\theta| < 1$.

b) This case can be reduced to the previous one. In fact, the hypothesis implies that $|a_n z_0^n| = |(a_0 + \cdots + a_n z_0^n) - (a_0 + \cdots + a_{n-1} z_0^{n-1})| \leq 2Kn^k$.

3. For the first two series we can state only that $R \geq \min(r, r')$, i.e. that $R \geq (r + r' - |r - r'|)/2$. In fact R may take on any value consistent with this inequality. If $a_n = 1$, $a_n' = -1 + 1/n!$, for instance, then $r = r' = 1$ and $R = +\infty$ (for the first series).

For $\Sigma a_n a_n' z^n$ we have $R \geq rr'$. In fact, for every $\epsilon > 0$ and for all sufficiently large n

$$|a_n|^{1/n} \leq \frac{1}{r} + \epsilon, \qquad |a_n'|^{1/n} \leq \frac{1}{r'} + \epsilon$$

so that $|a_n \cdot a_n'|^{1/n} \leq 1/rr' + \epsilon'$, ϵ' being a number which converges to 0 together with ϵ. (Give an example of series with $R > rr'$.)

For $\Sigma (a_n/a_n') z^n$ we have $R \leq r/r'$. Proof. Let ϵ be an arbitrary positive number. The inequality $|a_n|^{1/n} > 1/r - \epsilon$ holds for infinitely many n, the inequality $|a_n'|^{1/n} < 1/r' + \epsilon$ for all but a finite number of n. If both inequalities hold, then

$$\left| \frac{a_n}{a_n'} \right|^{1/n} > \frac{1/r - \epsilon}{1/r' + \epsilon} = \frac{r'}{r} - \epsilon'.$$

Thus $\overline{\lim} \mid a_n/a_n' \mid^{1/n} \geq r'/r$. (Give an example of series with $R < r/r'$.)

4. The first two series have the same radius of convergence as $\Sigma a_n z^n$ (since $n^{1/n} \to 1$). The third series is everywhere convergent ($n!^{1/n} \to +\infty$ and $\mid a_n \mid^{1/n} < 1/r + \epsilon$ for sufficiently large n, so that $\lim \mid a_n/n! \mid^{1/n} = 0$). The last series diverges for $z \neq 0$ if $r < +\infty$. (In this case $\mid a_n \mid^{1/n} > 1/r - \epsilon$ for infinitely many n.) No general statement can be made concerning the case $r = +\infty$.

5. The series diverges at $z = 1$. It converges at all other points of the unit circle. *Proof.* If $\mid z \mid = 1$, $z \neq 1$ then the partial sums of Σz^n are bounded at z, since

$$\mid 1 + z + \cdots + z^n \mid = \left| \frac{1 - z^{n+1}}{1 - z} \right| \leq \frac{2}{\mid 1 - z \mid}.$$

The coefficients $1/n$ decrease monotonically to 0. These two statements imply the convergence of $\Sigma z^n/n$ (see §3, problem 13a).

6. The series $\Sigma \mid a_n z_0^n \mid = \Sigma \mid a_n \mid r^n$ converges. For any fixed $\epsilon > 0$ we can find a n_0 such that

$$\mid a_{n+1} \mid r^{n+1} + \mid a_{n+2} \mid r^{n+2} + \cdots + \mid a_{n+p} \mid r^{n+p} < \epsilon$$

whenever $n > n_0$, $p \geq 1$. Thus

$$\mid a_{n+1} z^{n+1} \mid + \mid a_{n+2} z^{n+2} \mid + \cdots + \mid a_{n+p} z^{n+p} \mid < \epsilon$$

whenever $\mid z \mid \leq r$, $n > n_0$, $p \geq 1$. This proves the assertion.

7. The result of problem 5 implies that our series converges at all points of the unit circle, except at those points z for which $z^p = 1$, i.e. except at the vertices

of an inscribed regular p-gon with one vertex at $z = 1$.

8. a) The assertion follows from the fact that $| a | < 1$ for sufficiently large n.

b) The assertion follows from the result of §3, problem 13a (cf. problem 5). At $z = 1$ the series may either converge or diverge.

9. Use the result of §3, problem 13c.

10. No general statement can be made. (It is true that the series $\Sigma(a_n r^n)z^n$ has the radius of convergence 1, but the coefficients $(a_n r^n)$ may or may not approach 0.)

11. $R \geq \min (r, \rho)$. In fact, if $R \leq r$, $R \leq \rho$ and $| z | < R$, then the series $\Sigma a_n z^n$ and $\Sigma b_n z^n$ converge absolutely and so does $\Sigma c_n z^n$. Examples show that R may be greater than the smaller of the two numbers r and ρ or equal to this smaller number. (Consider the cases: $\Sigma a_n z^n \equiv \Sigma b_n z^n \equiv \Sigma z^n$; $\Sigma a_n z^n \equiv \Sigma z^n$, $\Sigma b_n z^n \equiv (1 - z)e^z$.)

12. We make use of the following theorem on infinite series. Given infinitely many absolutely convergent series

$$c_{00} + c_{01} + c_{02} + \cdots + c_{0n} + \cdots = C_0 ,$$

$$c_{10} + c_{11} + c_{12} + \cdots + c_{1n} + \cdots = C_1 ,$$

$$\dots\dots\dots\dots\dots\dots\dots\dots\dots\dots\dots\dots$$

set

$$| c_{k1} | + | c_{k2} | + \cdots + | c_{kn} | + \cdots = \gamma_k ,$$

$$(k = 0, 1, \cdots).$$

If $\Sigma\gamma_k$ converges, then all series

$$c_{0n} + c_{1n} + c_{2n} + \cdots + c_{kn} + \cdots = C'_n$$

converge, and

$$\sum_{k=0}^{\infty} C_k = \sum_{n=0}^{\infty} C'_n \,.$$

A detailed proof of this theorem will be found in the book *Theory and Applications of Infinite Series* by K. Knopp, Blackie and Son, Ltd., 1928.

Set

$$A_k(w - w_0)^k = A_k[(a_0 - w_0) + \sum a_n(z - z_0)^n]^k$$

$$= \sum_{n=0}^{\infty} a_n^{(k)}(z - z_0)^n, \qquad (k = 0, 1, \cdots).$$

These series converge absolutely for $|z - z_0| < r$. We must have

$$b_0 = A_0 + A_1(a_0 - w_0) + A_2(a_0 - w_0)^2 + \cdots$$

so that $|a_0 - w_0|$ should not exceed R. If $|a_0 - w_0| < R$, however, then

$$(!) \qquad |a_0 - w_0| + |a_1(z - z_0)| + |a_2(z - z_0)^2|$$
$$+ \cdots < R$$

for sufficiently small values of $|z - z_0|$.

We choose a R_1 such that $|a_0 - w_0| < R_1 < R$ and a ρ, $0 < \rho < r$, such that the right hand side in $(!)$ remains $\leq R_1$ for $|z - z_0| \leq \rho$. For $|z - z_0| \leq \rho$ our expansion of W is legitimate and yields $W = g(f(z))$. In fact, we may apply the theorem stated above to the case $c_{kn} = a_n^{(k)}(z - z_0)^n$, $|z - z_0| \leq \rho$, since we have $\gamma_k \leq |A_k| \cdot R_1^k$ so that $\Sigma \gamma_k$ converges

To give a better estimate of the radius of convergence of $\Sigma b_n(z - z_0)^n$ we would need more information concerning the function-theoretical character of the functions $f(z)$ and $g(w)$.

13. We have

$$W = \frac{1}{a_0 + a_1 z + \cdots} = \frac{1}{a_0} \cdot \frac{1}{1 + [(a_1/a_0)z + \cdots]}$$

$$= \frac{1}{a_0} \left\{ 1 - \left[\frac{a_1}{a_0} z + \cdots \right] + \left[\frac{a_1}{a_0} z + \cdots \right]^2 \right.$$

$$\left. - + \cdots \right\}.$$

According to the result of problem 12 we may obtain the expansion of W by computing the powers of $[(a_1/a_0)z + (a_2/a_0)z^2 + \cdots]$ and collecting terms. The resulting series will certainly converge for $|z| < \rho$, ρ being the uniquely determined positive number such that

$$|a_1| \cdot \rho + |a_2| \rho^2 + \cdots = |a_0|.$$

Set $W = b_0 + b_1 z + \cdots$. We have that

$$(a_0 + a_1 z + a_2 z^2 + \cdots)(b_0 + b_1 z + b_2 z^2 + \cdots) \equiv 1$$

so that

$$a_0 b_0 = 1$$

$$a_0 b_n + a_1 b_{n-1} + \cdots + a_n b_0 = 0, \qquad n = 1, 2, \cdots.$$

From these equations we obtain at once the recursion formulas for the coefficients b_n.

14. If $|z_1| = 1$, then the power series

$$\sum \frac{1}{n}\left(\frac{z}{z_1}\right)^n$$

converges for $|z| = 1$, $z \neq z_1$, diverges at $z = z_1$ (see problem 5). Let the p preassigned points be z_1, z_2, $\cdots$, z_p. The series

$$\sum_{n=1}^{\infty} \frac{1}{n}\left(\frac{1}{z_1^n} + \frac{1}{z_2^n} + \cdots + \frac{1}{z_p^n}\right)z^n$$

possesses the required property.

15. Yes. The first example of such a series was given in 1912 by W. Sierpiński. A clear presentation will be found in the book by E. Landau, *Darstellung und Begründung einiger neuerer Engebuisse der Funktionentheorie*, 2nd edition, Berlin, 1929, p. 71. The proof is too complicated to be reproduced here.

CHAPTER III

FUNCTIONS OF A COMPLEX VARIABLE

§5. Limits of Functions. Continuity and Differentiability

1. a) $f(z)$ is continuous at $z = 0$ and at all points z for which $|z|$ is irrational. At all other points $f(z)$ is discontinuous. *Proof.* Assume that $|z_0|$ is irrational. For a given $\epsilon > 0$ there exists a sufficiently small circle with center at z_0 such that for all points z within this circle and with a rational $|z| = p/p$, $q > 1/\epsilon$. It follows that for these points z $|f(z_0) - f(z)| < \epsilon$. A similar proof holds for $z = 0$. Now let $|z_0| = p_0/q_0$ be rational. Choose an ϵ such that $0 < \epsilon < 1/q_0$. There are points z arbitrarily close to z_0 for which $|f(z) - f(z_0)| > \epsilon$, in fact, all those for which $|z|$ is irrational.

2. The assertion is an almost immediate consequence of the definition of continuity. For any given $\epsilon > 0$ there exists a $\delta > 0$ such that $|f(z) - f(\zeta)| < \epsilon$ whenever $|z - \zeta| < \delta$. Since $z_n \to \zeta$ we will have $|z_n - \zeta| < \delta$ and hence $|w_n - f(\zeta)| < \epsilon$ provided $n > n_0$, n_0 being sufficiently large. This shows that $w_n \to f(\zeta)$.

3. Assume $f(z)$ to be discontinuous at ζ. Then there exists a $\epsilon_0 > 0$ such that every neighborhood of ζ contains at least one point z for which $|f(z) - f(\zeta)| \geq \epsilon_0$. Let z_n be such a point in the circular neighborhood of ζ with radius $1/n$, $n = 1, 2, \cdots$. Then $|z_n - \zeta| < 1/n$ and $|f(z_n) - f(\zeta)| \geq \epsilon_0$; i.e., $z_n \to \zeta$ but not $f(z_n) \to f(\zeta)$. This contradicts the hypothesis.

4. The function $f(z) = 1/(1 - z)$ is continuous for $|z| < 1$ but not uniformly continuous. In fact, let for each point z, $|z| < 1$, ρ_z denote the largest number such that $|f(z') - f(z)| < 1$ whenever $|z' - z| < \rho_z$ and $|z'| < 1$. The lower limit of the numbers ρ_z is 0, since for $z = 1/n$, $\rho_z = [1 - 1/(n + 1)] - (1 - 1/n) < 1/n^2$.

5. Since $z \neq 0$ in the domain considered, the function is continuous. The continuity is uniform. In fact, set $f(0) = 0$. Then $f(z)$ becomes continuous at $z = 0$ (proof?). It is even continuous in the domain $|z| < 2$, hence also in the *closed* domain $|z| \leq 1$. It follows (KI 6, p. 25) that $f(z)$ is uniformly continuous in $|z| \leq 1$ and *a fortiori* in the subdomain $0 < |z| < 1$.

6. The numerators and denominators of all five functions are continuous in the whole plane. The denominator of f_1 never vanishes. The denominators of the other four functions vanish only at $z \neq 0$. It follows that f_1 is continuous in the entire plane and $f_2 - f_5$ are continuous everywhere, except perhaps at $z = 0$. The functions f_2 and f_4 *are* discontinuous at the origin, for both have the limit 1 when z approaches the origin from the right along the x-axis, and the limit 0 when z approaches the origin from above along the y-axis. The functions f_3 and f_5 are continuous at the origin. In fact, since $|\Re(z)| \leq |z|$ we have

$$|f_3(z)| \leq |\Re(z)| \leq |z|, \quad |f_5(z)| \leq \Re(z^2) \leq |z|^2.$$

If $z \to 0$, then $f_3(z) \to 0 = f_3(0)$ and $f_5(z) \to 0 = f_5(0)$.

7. The answers are the same as in 6, the proof similar.

8. Uniform continuity of $f(z)$ implies that to every $\epsilon > 0$ there exists a $\delta > 0$ such that $|f(z') - f(z)| <$

ϵ whenever $|z' - z| < \delta$ (and $|z'| < 1$, $|z| < 1$).
Let (z_n) be a sequence satisfying the hypotheses of the
theorem. There exists a number n_0 such that $|z_n - \zeta| <$
$\delta/2$ for $n > n_0$. Thus, for $n > n_0$, $m > n_0$, we have
that $|z_n - z_m| < \delta$ so that $|f(z_n) - f(z_m)| < \epsilon$. It
follows that $w = \lim_{n \to \infty} f(z_n)$ exist. To show that w depends only upon ζ, consider another sequence (z'_n) satisfying the conditions of the theorem. $w' = \lim_{n \to \infty} f(z'_n)$
exists. Since the sequence $z_1, z'_1, z_2, z'_2, \cdots$ satisfies the
hypotheses, the sequence $f(z_1), f(z'_1), f(z_2), f(z'_2), \cdots$
converges to a limit W. But since the limit of a convergent sequence is identical with that of any subsequence, $W = w = w'$.

9. Let ζ' and ζ'' be two boundary points, $f(\zeta')$ and
$f(\zeta'')$ the corresponding limit values of $f(z)$. Let $\epsilon > 0$
be given and δ determined as in problem 8. We shall
show that $|f(\zeta') - f(\zeta'')| \leq \epsilon$ provided that
$|\zeta' - \zeta''| < \delta/2$. In fact, if (z'_n) and (z''_n) are sequences
of interior points converging to ζ' and ζ'', then
$|\zeta' - \zeta''| < \delta/2$ implies that $|z'_n - z''_n| < \delta$ for all
sufficiently large n. In this case $|f(z'_n) - f(z''_n)| < \epsilon$,
and since $f(z'_n) \to f(\zeta)$, $f(z''_n) \to f(\zeta'')$ also $|f(\zeta') -
f(\zeta'')| \leq \epsilon$. This proves the (uniform) continuity of
the boundary values. If these boundary values are used
to define the function $f(z)$ for $|z| = 1$, the resulting
function is uniformly continuous in $|z| \leq 1$.

10. a) The function $f(z)$ in problem 1a is nowhere
differentiable. It could possess a derivative only at
$z_0 = 0$ or at a point z_0 for which $|z_0|$ is irrational, since
at all other points the function is not even continuous.
In every neighborhood of $z_0 = 0$ there are points with
with $f(z) = 0$, and also points with $f(z) = z$ (the points

$z = 1/n$). It follows that for $z_0 = 0$ the difference quotient

$$D(z, z_0) = \frac{f(z) - f(z_0)}{z - z_0}$$

takes on the value 0 and 1 in every neighborhood of z_0. $f'(0)$ does not exist. If $z_0 \neq 0$ and $|z_0|$ is irrational, then $D(z, z_0) = 0$ for $|z| = |z_0|$, and $D(z, z_0) = 1/[q(z - z_0)]$ for $|z| = p/q$. In particular, if $|z| = p/q$ and z, z_0 and 0 are colinear, then

$$|D(z, z_0)| = \frac{1}{|p - q|z_0||}.$$

It is known, however, that for any irrational number γ there exist positive integers p and q for which $|p - q\gamma|$ is arbitrarily small. It follows that $|D(z, z_0)|$ becomes both 0 and arbitrarily large in every neighborhood of z_0. Hence $f'(z_0)$ does not exist.

b) The function in 1b is differentiable at all points of the imaginary axis (except for $z_0 = 0$ where it is discontinuous) and at no other point. Proof: if $z_0 \neq 0$ and $z \to z_0$ along a straight line through the origin, $D(z, z_0) \to 0$; if $z \to z_0$ along the circle $|z| = |z_0|$, $D(z, z_0) \to \cos \theta_0 = \operatorname{am} z_0$. Hence the function can be differentiable only if $\cos \theta_0 = 0$; i.e., if $\Re(z_0) = 0$, in which case $f'(z_0)$ must vanish. The function actually is differentiable at such a point. In fact, set $\operatorname{am} z = \theta = \pi/2 - \eta$. For z sufficiently close to z_0, $|z - z_0| \geq |z_0| \cdot \sin \eta$, so that

$$|D(z_1 z_0)| \leq \frac{1 - \cos \eta}{|z_0| \cdot \sin \eta} < \frac{\tan (\eta/2)}{|z_0|} < \epsilon$$

provided $|z_0 - z|$ is sufficiently small.

c) The function in 5 is not differentiable at $z_0 \neq 0$, since $D(z, z_0) = 0$ for $|z| = |z_0|$ and $D(z, z_0) \to |z_0|^{-2} e^{-1/|z_0|} \cdot e^{-i \, \text{am} \, z_0} \neq 0$ when $z \to z_0$ along a straight line through the origin. If one sets $f(0) = 0$, the function becomes differentiable at $z_0 = 0$, since $|D(z, 0)| = |z|^{-1} e^{-1/|z|} \to 0$ for $z \to 0$.

11. The proof follows by direct computation. The computation may be avoided by noting that $\phi_1 + i\psi_1 = \phi_2 + i\psi_2 = e^f$, where $f = \phi + i\psi$ and using the fact that a regular analytic function of a regular analytic function is itself regular analytic.

12. Since any two points of $\mathfrak{G}$ may be joined by a polygonal path it suffices to show that $f(z)$ possesses the same value at the end-points of any straight segment s situated in G. Let l be the length of s, ζ' and ζ'' its endpoints, $\epsilon > 0$ an arbitrarily given number. By hypothesis each point z' of s possesses a neighborhood such that for any point z'' of this neighborhood

$$\left| \frac{f(z'') - f(z')}{z'' - z'} \right| < \frac{\epsilon}{2l}.$$

Applying the Heine-Borel theorem (KI, 3, p. 9) to the closed set s is follows that a finite number of such neighborhoods covers s. Hence we can find a sequence of points $z_0 = \zeta'$, z_1, z_2, $\cdots$, $z_p = \zeta''$ on s such that two consecutive points lie in one such neighborhood. We have then, for $\nu = 1, 2, \cdots, p$,

$$|f(z_\nu) - f(z_{\nu-1})| < \frac{\epsilon}{l} |z - z_{-1}|.$$

Adding these p inequalities we get

$$|f(\zeta'') - f(\zeta')| < \epsilon.$$

Since ϵ was arbitrary it follows that $f(\zeta') = f(\zeta'')$. (A shorter but less direct proof follows from theorem 1, KI 20, p. 79.)

13. Set $f(z) = \phi(x, y) + i\psi(x, y)$, $\zeta = \xi + i\eta$. ϕ and ψ have continuous partial derivatives satisfying the Cauchy-Riemann equations. We must prove that

$$\frac{\phi(x_n', y_n') + i\psi(x_n', y_n') - \phi(x_n, y_n) - i\psi(x_n, y_n)}{(x_n' - x_n) + i(y_n' - y_n)}$$

$$\to \phi_x(\xi, \eta) + i\psi_x(\xi, \eta)$$

We prove this for the real parts of these expressions. The real part of the fraction is

$$\{[\phi(x_n', y_n') - \phi(x_n, y_n)](x_n' - x_n) + [\psi(x_n', y_n')$$

$$- \psi(x_n, y_n)](y_n' - y_n)\}/\{(x_n' - x_n)^2 + (y_n' - y_n)^2\}.$$

Applying to the expressions in the brackets the mean-value theorem and noting that the partial derivatives are continuous at (ξ, η), we see that the two terms of the numerator are equal to

$$\{[\phi_x(\xi, \eta) + \alpha](x_n' - x_n) + [\phi_y(\xi, \eta)$$

$$+ \alpha'](y_n' - y_n)\}(x_n' - x_n)$$

and

$$\{[\psi_x(\xi, \eta) + \beta](x_n' - x_n) + [\psi_y(\xi, \eta)$$

$$+ \beta](y_n' - y_n)\}(y_n' - y_n)$$

respectively, α, α', β, β' being quantities which approach 0 when $(x_n, y_n) \to (\xi, \eta)$, $(x_n', y_n') \to (\xi, \eta)$. Since $\phi_x = \psi_y$, $\phi_y = -\psi_x$ the real part of our difference quotient may be written as

$$\phi_x(\xi, \eta) + \{\alpha(x_n' - x_n)^2 + (\alpha' + \beta)(x_n' - x_n)(y_n' - y_n)$$

$$+ \beta'(y_n' - y_n)^2\}/\{(x_n' - x_n)^2 + (y_n' - y_n)^2\}.$$

This fraction $\to 0$ as z_n and $z_n' \to \zeta$, for it may be written as a sum of three products of two terms each, one term converging to 0 and the other remaining <1 in absolute value. Thus the real part of the difference quotient $\to \phi_x(\xi, \eta)$. In a similar way we can show that the imaginary part $\to \psi_x(\xi, \eta)$.

14. Assume the statement to be wrong. Then there exist an $\epsilon_0 > 0$ and two sequences (z_n) and (z_n') in $|z - \zeta| \leq \rho'$ such that $|z_n' - z_n| \to 0$ but

$$\left| \frac{f(z_n') - f(z_n)}{z_n' - z_n} - f'(z_n) \right| \geq \epsilon_0.$$

Let Z be a limit point of (z_n). There exists a subsequence (ζ_n) of (z_n) such that $\zeta_n \to Z$. The corresponding subsequence (ζ_n') of (z_n) also converges to Z. We have

$$\left| \frac{f(\zeta_n') - f(\zeta_n)}{\zeta_n' - \zeta_n} - f'(\zeta_n) \right| \geq \epsilon_0$$

for all n. By virtue of the result of the preceding problem, and since $f'(z)$ is assumed to be continuous, the above expression converges to $|f'(Z) - f'(Z)| = 0$. Thus we arrive at a contradiction.

§6. Simple Properties of the Elementary Functions

1. For $0 < |z| < 1$

$$|e^z - 1| \leq |z| \left\{ 1 + \frac{1}{2!} + \frac{1}{3!} + \cdots \right\}$$

$$= (e - 1) |z| < \frac{7}{4} |z|$$

and

$$|e^z - 1| > |z| \left\{ 1 - \frac{1}{2!} - \frac{1}{3!} - \cdots \right\}$$

$$= (3 - e) |z| > \frac{1}{4} |z|.$$

2. We have $|e^z - 1| \leq |z| + |z|^2/2! + \cdots$. This series is equal to $e^{|z|} - 1$, and also to

$$|z| \left(1 + \frac{|z|}{2!} + \frac{|z|^2}{3!} + \cdots \right) \leq |z| \left(1 + \frac{|z|}{1!} \right.$$

$$\left. + \frac{|z|^2}{2!} + \cdots \right) = |z| e^{|z|}.$$

3. a) Consider the function $f(x) = e^x - (1 + x)$. $f(0) = 0$ and $f'(x) = e^x - 1$. Thus $f'(x) > 0$ for $x > 0$, and $f'(x) < 0$ for $x < 0$. This shows that $f(x) > 0$ for $x \neq 0$.

   b) is equivalent to a) for $1 - x > 0$. (Take the reciprocals of both sides and set $-x = x'$.)

   c) The first inequality is equivalent to a) (and

holds for all real x), the second to b) (replace x by $-x$).

d) is equivalent to c).

e) follows from b) by replacing x by $x/(1 + x)$.

f) follows directly from the defining power series, all terms being positive.

g) Replace in a) and e) x by x/y.

h) $e^{-x} = 1 - \dfrac{x}{2}(2 - x) - \dfrac{x^3}{4!}(4 - x) - \cdots$

$$< 1 - \frac{x}{2}(2 - x) = 1 - \frac{x}{2}.$$

4. We can repeat, almost word by word, the classical proof. Assume that $|z| \leq r$ and choose an $\epsilon > 0$. Let p be a fixed positive integer such that

$$\frac{r^{p+1}}{(p + 1)!} + \frac{r^{p+2}}{(p + 2)!} + \cdots < \frac{\epsilon}{2}.$$

For $n > 2$ we have

$$\left(1 + \frac{z}{n}\right)^n = 1 + z + \frac{1}{2!}\left(1 - \frac{1}{n}\right)z^2 + \cdots$$

$$+ \frac{1}{\nu!}\left[\left(1 - \frac{1}{n}\right) \cdots \left(1 - \frac{\nu - 1}{n}\right)\right]z^\nu + \cdots .$$

The coefficients of z^ν are positive numbers not exceeding 1 and are smaller than the corresponding coefficients in the series $1 + z + z^2/2! + \cdots$. It follows that

$$\left| \sum_{\nu=0}^{\infty} \frac{z^\nu}{\nu!} - \left(1 - \frac{z}{n}\right)^n \right| \leqq \frac{1}{2!}\left[1 - \left(1 + \frac{1}{n}\right)\right]r^2 + \cdots$$

$$+ \frac{1}{p!}\left[1 - \left(1 - \frac{1}{n}\right)\left(1 - \frac{2}{n}\right)\right.$$

$$\left.\cdots \left(1 - \frac{p-1}{n}\right)\right] r^p + \frac{\epsilon}{2}.$$

Since p is fixed the right hand side approaches $0 + \epsilon/2 = \epsilon/2$ when $n \to +\infty$. It follows that the left hand side is less than ϵ for $n > n_0 > p$. Thus $\sum_{\nu=0}^{\infty} z^\nu/\nu! = \lim_{n \to +\infty} (1 + z/n)^n$ as asserted. Note that we proved that the convergence $(1 + z/n)^n \to e^z$ is uniform in every bounded domain.

5. Let r be so large that $|z_n| < r$ for $n = 1, 2, \cdots$, and $|\zeta| \leq r$. For every given $\epsilon > 0$, there exists an n_0 such that

$$\left|\left(1 + \frac{z}{n}\right)^n - e^z\right| < \frac{\epsilon}{2}$$

for $|z| \leq r$ and $n > n_0$. (This was proved in the answer to problem 4.) In particular, whenever $n > n_0$,

$$\left|\left(1 + \frac{z_n}{n}\right)^n - e^{z_n}\right| < \frac{\epsilon}{2}.$$

Since e^z is continuous and $z_n \to \zeta$ we may choose a $m > n_0$ such that

$$|e^{z_n} - e^\zeta| < \frac{\epsilon}{2}$$

whenever $n > m$. For $n > m$ we have

$$\left|\left(1 + \frac{z}{n}\right)^n - e^\zeta\right| < \epsilon.$$

This proves the assertion.

6. If $e^{x+iy} = e^x (\cos y + i \sin y) = +1$, then $\sin y = 0$ so that $y = n\pi$ and $\cos y = (-1)^n$. Hence $n = 2k$, for otherwise e^{x+iy} would be negative. Since $e^x = 1$ if and only if $x = 0$ the only solutions are $z = 2k\pi i$.

7. $\displaystyle\sum_{n=0}^{\infty} \frac{z_1^n}{n!} \cdot \sum_{n=0}^{\infty} \frac{z_2^n}{n!}$

$$= \sum_{n=0}^{\infty} \left(\frac{z_1^n}{n!} + \frac{z_1^{n-1} \cdot z_2}{(n-1)!1!} + \cdots \right.$$

$$\left. + \frac{z_1 \cdot z_2^{n-1}}{1!(n-1)!} + \frac{z_2^n}{n!} \right)$$

$$= \sum_{n=0}^{\infty} \frac{1}{n!} \left[\binom{n}{0} z_1^n + \binom{n}{1} z_1^{n-1} z_2 + \cdots + \binom{n}{n} z_2^n \right]$$

$$= \sum_{n=0}^{\infty} \frac{(z_1 + z_2)^n}{n!} = e^{z_1 + z_2}.$$

8. We give only a brief sketch of the proof. Consider first the series

$$C(y) = 1 - \frac{y^2}{2!} + \frac{y^4}{4!} - + \cdots ,$$

$$S(y) = y - \frac{y^3}{3!} + \frac{y^5}{5!} - + \cdots$$

for real values of y. $C(0) = 1 > 0$, $C(2) < 0$ so that the equation $C(y) = 0$ has at least one root between 0 and 2, in fact exactly one, since $C'(y) = -S(y) < 0$ for $0 < y < 2$. Call this root $\pi/2$. Next prove by direct computations (cf. problems 7 and 13) the addition-theorems: $C(y_1 + y_2) = C(y_1)C(y_2) - S(y_1)S(y_2)$, $S(y_1 + y_2) = S(y_1)C(y_2) + S(y_2)C(y_1)$. They imply that $C^2(y) + S^2(y) = 1$, $C(2y) = 2C^2(y) - 1$, $S(2y) = 2S(y)C(y)$. Hence $C(\pi) = -1$, $S(\pi) = 0$, $S(2\pi) = 0$, so that $C(y + 2\pi) = C(y)$, $S(y + 2\pi) = S(y)$. Thus 2π is a (*primitive*) period of $C(y)$ and $S(y)$. According to problem 7

$$e^{z+2\pi i} = e^z \cdot e^{2\pi i} = e^z(C(2\pi) + iS(2\pi)) = e^z,$$

so that $2\pi i$ is a period of e^z.

9. Set am $z = \phi$. For $-\pi/2 < \phi < \pi/2$, $\Re(z) = x \to +\infty$ so that $e^z \to \infty$, since $|e^z| = e^x$. If $\pi/2 < \phi < 3\pi/2$, $e^z \to 0$ since $|e^z| = e^x \to 0$. If $\phi = \pm\pi 2$, e^z has no limit since $\Re(e^z) = \cos y$ oscillates between $+1$ and -1.

10. $z + e^z \to \infty$ for all directions. In order to prove it apply the following estimates:

$$|z + e^z| \geq |e^z| - |z| \qquad \text{for } -\frac{\pi}{2} < \phi < \frac{\pi}{2},$$

$$|z + e^z| \geq |z| - |e^z| \qquad \text{for } \frac{\pi}{2} < \phi < \frac{3\pi}{2},$$

$$|z + e^z| \geq |z| - 1 \qquad \text{for } \phi = \pm\frac{\pi}{2}.$$

11. The limits are 0 and ∞ along the branches for which the x-axis is an asymptote. There are no limits along the other branches.

12. The limits exist along both branches. Their values are 0 and ∞.

13. The computation is very similar to the one performed for problem 7.

14. $e^z = e^{x+iy} = e^x (\cos y + i \sin y)$ is real if and only if $\sin y = 0$, i.e. for $y = \Im(z) = k\pi (k = 0, \pm 1, \pm 2, \cdots)$. e^z is positive (negative) if k is even (odd). $\sin z = \sin (x + iy) = \sin x \cdot (e^y + e^{-y})/2 + \cos x \cdot (e^y + e^{-y})/2i$ is real if and only if either $y = 0$ (and z itself real) or $\cos x = 0$, i.e. $x = (2k + 1) \pi/2$ ($k = 0, \pm 1, \pm 2, \cdots$). Along the straight lines $\Re(z) = (2k + 1) \pi/2$, $\sin z$ has the same sign as $(-1)^k$. The answer to the question concerning $\cos z$ is obtained at once by noting that $\cos z = \sin (z + \pi/2)$.

15. Set $z = x + iy$ and apply the addition theorems. Thus

$$e^{2+i} = e^2 (\cos 1 + i \sin 1)$$

and since the angle $1 = 57° 17' 44.8''$, tables yield the result

$$e^{2+i} \approx 3.992 + i\, 6.218.$$

Similarly,

$$\cos (5 - i) = \cos 5 \cdot \frac{e + e^{-1}}{2} + i \sin 5 \cdot \frac{e - e^{-1}}{2}$$

$$\approx 0.438 - i\, 1.127,$$

$$\sin (1 - 5i) = \sin 1 \cdot \frac{e^5 + e^{-5}}{2} - i \cos 1 \cdot \frac{e^5 - e^{-5}}{2}$$

$$\approx 62.45 - i\, 40.09.$$

16. Since

$$\sin z_2 - \sin z_1 = 2 \cos \frac{z_2 + z_1}{2} \sin \frac{z_2 - z_1}{2}$$

the right hand side vanishes if and only if either $(z_2 - z_1)/2 = k\pi$ or $(z_2 + z_1)/2 = (2k + 1) \pi/2$ ($k = 0, \pm 1, \pm 2, \cdots$), for $\sin z$ and $\cos z$ have no zeros outside the real axis. Thus $\sin z_2 = \sin z_1$ if and only if either $z_2 = z_1 + 2k\pi$ or $z_2 = \pi - z_1 + 2k\pi$. It follows that all solutions of the equation $\sin z = c$ can be obtained from one solution z_1 with $-\pi < \Re(z_1) < \pi$. Now if $\sin z = (e^{iz} - e^{-iz})/2i = c$, i.e. $e^{iz} - e^{-iz} = 2ic$, then $e^{iz} = ic + (1 - c^2)^{1/2}$. Set $ic + (1 - c^2)^{1/2} = Ce^{i\gamma}$, C and γ being real. Then $e^{iz} = Ce^{i\gamma}$ or $iz = \log C + i\gamma$. Thus *one* solution is given by

$$z_1 = \gamma - i \log C.$$

(One may use any branch of $(1 - c^2)^{1/2}$, but once this value is chosen it should be kept fixed.)

For $c = 1000$ one gets

$$ic + (1 - c^2)^{1/2} = i (1000 + 999999^{1/2})$$

so that

$$\gamma = \frac{\pi}{2}, \qquad C = 1000 + 999999^{1/2}$$

and

$$z_1 \approx \frac{\pi}{2} - i \, 7.601.$$

For $c = 5i$, $z_1 \approx i \cdot 2.312$; for $c = 1 - i$, $z_1 \approx 0.6663 - i \cdot 1.0613$.

The procedure for solving the equation $\cos z = c$ is

quite similar. One solution is given by $z_1 = \gamma - i \log C$ where $Ce^{i\gamma} = c + (c^2 - 1)^{1/2}$. For $c = 2i$, $4 + 3i$, 5, one gets

$$z_1 \approx \frac{\pi}{2} - i \cdot 1.444,$$

$$z_1 \approx 0.650 - i \cdot 2.300,$$

$$z_1 \approx i \cdot 2.292.$$

17. We must have

$$(a_0 + a_1 z + \cdots)\left(1 - \frac{z^2}{2!} + \frac{z^4}{4!} - + \cdots\right)$$

$$= z - \frac{z^3}{3!} + \frac{z^5}{5!} - + \cdots .$$

Multiplication and comparison of coefficients leads to the equations

$$a_0 = 0 \qquad\qquad -\frac{a_1}{2} + a_3 = -\frac{1}{6}$$

$$a_1 = 1 \qquad\qquad \frac{a_0}{24} - \frac{a_2}{2} + a_4 = 0$$

$$-\frac{a_0}{2} + a_2 = 0 \qquad \frac{a_1}{24} - \frac{a_3}{2} + a_5 = \frac{1}{120}$$

which yield the values $a_0 = 0$, $a_1 = 1$, $a_2 = 0$, $a_3 = 1/3$, $a_4 = 0$, $a_5 = 2/15$.

18. The relation $\cos^2 z + \sin^2 z = 1$ implies only

that $\cos z$ equals to *one* of the two values of $(1 - \sin^2 z)^{1/2}$.

19. Since

$$\log z = \log [1 + (z - 1)] = \frac{z - 1}{1} - \frac{(z - 1)^2}{2}$$

$$+ \frac{(z - 1)^3}{3} - + \cdots$$

we have

$$| \log z | \leqq \rho + \frac{\rho^2}{2} + \frac{\rho^3}{3} + \cdots \leqq \rho + \rho^2 + \rho^3 + \cdots$$

$$= \frac{\rho}{1 - \rho}.$$

Equality holds only for $\rho = 0$, $z = 1$.

20. We need a domain in which am sin z can be defined as a single-valued function. We choose the domain G as the whole plane except the positive real numbers ≥ 1 and the negative real numbers ≤ -1. For every z of G there exists exactly one value w such that $- \pi/2 < \Re(w) < \pi/2$ and $\sin w = z$. In this way we define $w = \arccos \sin z$. Since $\sin w = z$ we have $\cos w dw = dz$ so that

$$\frac{dw}{dz} = \frac{1}{\cos w}.$$

The right hand side is equal to one of the values of

$$\frac{1}{(1 - \sin^2 w)^{1/2}} = \frac{1}{(1 - z^2)^{1/2}}.$$

But $(1 - z^2)^{1/2}$ separates in G in two single-valued branches. Since dw/dz is continuous in G and takes on the value $+1$ at $z = 0$ we must choose that branch of $(1 - z^2)^{1/2}$ which equals to $+1$ at $z = 0$. Now we may extend the validity of our formula to points of the boundary of G, which is to be considered as having two banks (Proof?).

21. a) $w = \arcsin z$ is a solution of the equation $\sin w = z$. Hence (cf. problem 16)

$$e^{iw} = iz + (1 - z^2)^{1/2}$$

or

$$w = -i \log \{iz + (1 - z^2)^{1/2}\}.$$

b) In the same way we obtain from $\tan w = z$, i.e. from $(e^{iw} - e^{-iw})/(e^{iw} + e^{-iw}) = iz$,

$$w = \arctan z = \frac{i}{2} \log \frac{1 - iz}{1 + iz}.$$

22. $i^i = e^{i \log i} = e^{i[(\pi i)/2 + 2k\pi i]} = e^{-\pi/2 + 2k\pi}.$

Thus this power has only real values, the principal value being $e^{-\pi/2}$. The power $a^b = e^{b(\log a + 2k\pi i)}$ is single-valued if and only if b is an integer ($\gtreqless 0$), has q determinations if $b = p/q$, p and q being relatively prime integers, $p > 0$, is infinitely many-valued otherwise.

23. In any simply connected domain not containing $z = 0$ the function z^a separates in infinitely many single-valued branches. For each branch

$$dz^a = az^a \frac{dz}{z}$$

provided the same determination is used on both sides. Proof: Since $z^a = e^{a \log z}$, $dz^a = e^{a \log z} \cdot a(1/z)$.

24. The proofs follow almost immediately from the definitions (by making use of $e^{z_1 + z_2} = e^{z_1} e^{z_2}$) or from the obvious relations: $\operatorname{sh} z = -i \sin iz$, $\operatorname{ch} z = \cos iz$.

25. Since $\operatorname{ch} z = \cos iz$, a rotation of the z-plane about the origin by $-90°$ takes the curves along which $\cos z$ is real into the curves along which $\operatorname{ch} z$ is. The curves along which $\operatorname{sh} z$ is real are obtained from the curves along which $\sin z$ is purely imaginary by the same rotation, since $\operatorname{sh} z = -i \sin iz$.

CHAPTER IV

INTEGRAL THEOREMS

§7. Integration in the Complex Domain

1. a) Set $z = it$, $t = -1 \cdots +1$. Then the integral is given by

$$I = i \int_{-1}^{+1} |t| \, dt = 2i \int_0^1 t \, dt = i.$$

b) $z = \cos t + i \sin t$, $t = -\pi/2 \cdots - 3\pi/2$. Hence

$$I = \int dz = [z]_{t=-\pi/2}^{t=-3\pi/2} = 2i.$$

c) $I = [z]_{t=-\pi/2}^{t=\pi/2} = 2i.$

2. a) $z = \cos t + i \sin t$, $t = 0 \cdots 2\pi$. Hence

$$I = \int_0^{2\pi} \cos t(-\sin t + i \cos t) \, dt$$

$$= \left[\frac{1}{4} \cos 2t + \frac{it}{2} + \frac{i}{4} \sin 2t \right]_0^{2\pi} = \pi i.$$

b) $z = z_1 + (z_2 - z_1)t$, $t = 0 \cdots 1$. Hence

$$I = \int_0^1 [x_1 + (x_2 - x_1)t](z_2 - z_1) \, dt$$

$$= (z_2 - z_1) \frac{x_1 + x_2}{2}$$

where $x_1 = \Re(z_1)$, $x_2 = \Re(z_2)$.

c) $z = z_0 + r(\cos t + i \sin t)$, $t = 0 \cdots 2\pi$. Hence

$$I = \int_0^{2\pi} (x_0 + r \cos t)\, dz$$

$$= x_0 [z]_{t=0}^{t=2\pi} + r \int_0^{2\pi} \cos t \cdot dz = r^2 \pi i$$

(according to a)).

3. a) Let the four vertices of the square be z_1, z_2, z_3, z_4, $|z_i - z_j| = 2\alpha$, $i \neq j$. For $m \neq -1$

$$I = \int_{-\alpha}^{+\alpha} (t - i\alpha)^m\, dt + \int_{-\alpha}^{+\alpha} (\alpha + it)^m \cdot i \cdot dt$$

$$+ \int_{+\alpha}^{-\alpha} (t + it)^m\, dt + \int_{+\alpha}^{-\alpha} (-\alpha + it)^m \cdot i \cdot dt$$

$$= \frac{1}{m+1} \left\{ [z^{m+1}]_{z_1}^{z_2} + [z^{m+1}]_{z_2}^{z_3} + [z^{m+1}]_{z_3}^{z_4} + [z^{m+1}]_{z_4}^{z_1} \right\}$$

$$= 0.$$

For $m = -1$

$$I = \int_{-\alpha}^{+\alpha} \frac{dt}{t - i\alpha} + \int_{-\alpha}^{+\alpha} \frac{i\,dt}{\alpha + it} - \int_{-\alpha}^{+\alpha} \frac{dt}{t + i\alpha}$$

$$- \int_{-\alpha}^{+\alpha} \frac{i\,dt}{-\alpha + it}$$

$$= 4i\alpha \int_{-\alpha}^{+\alpha} \frac{dt}{\alpha^2 + t^2} = 4i \cdot \left[\arctan \frac{t}{\alpha} \right]_{-\alpha}^{+\alpha} = 2\pi i.$$

b) $z = z_0 + \alpha \cos t + i\beta \sin t$, $t = 0 \cdots 2\pi$. Hence

$$I = \int_0^{2\pi} (\alpha \cos t + i\beta \sin t)^m (-\alpha \sin t + i\beta \cos t)\, dt.$$

If m is even, separate the integral into $\int_0^\pi$ and $\int_\pi^{2\pi}$. The substitution $t = \pi + t$ takes the second integral into $-\int_0^\pi$. Thus $I = 0$. For an odd m the computation is more complicated. But for $m = -1$

$$I = \int_0^{2\pi} \frac{-\alpha \sin t + i\beta \cos t}{\alpha \cos t + i\beta \sin t}\, dt$$

$$= -(\alpha^2 + \beta^2) \int_0^{2\pi} \frac{\sin t \cos t}{\alpha^2 \cos^2 t + \beta^2 \sin^2 t}\, dt$$

$$+ i\alpha\beta \int_0^{2\pi} \frac{dt}{\alpha^2 \cos^2 t + \beta^2 \sin^2 t} = 2\pi i.$$

4. In both cases set $z = \cos t + i \sin t$, $z^{1/2} = \cos t/2 + i \sin t/2$. In case a) integrate from 0 to π, in case b) from 0 to $-\pi$. This yields

a) $I = -2(1 - i)$,   b) $I = -2(1 + i)$.

5. $I_1 = \int_\rho^r \frac{e^{ix}}{x}\, dx = \int_\rho^r \frac{\cos x}{x}\, dx + i \int_\rho^r \frac{\sin x}{x}\, dx.$

These integrals cannot be expressed in closed form by elementary functions. The same is true for the following integrals.

$$I_2 = i \int_0^\pi e^{-r \sin t + ir \cos t}\, dt$$

$$= -\int_0^\pi e^{-r \sin t} \sin (r \cos t)\, dt$$

$$+ i \int_0^\pi e^{-r \sin t} \cos (r \cos t) \, dt.$$

$$I_3 = \int_{-r}^{-\rho} \frac{e^{ix}}{x} \, dx = -\int_\rho^r \frac{e^{-ix}}{dx} = -\int_\rho^r \frac{\cos x}{x} \, dx$$

$$+ i \int_\rho^r \frac{\sin x}{x} \, dx.$$

$$I_4 = i \int_{+\pi}^0 e^{-\rho \sin t + i\rho \cos t} \, dt$$

$$= -i \int_0^{+\pi} e^{-\rho \sin t + i\rho \cos t} \, dt.$$

6. a) $\ \left| I_2 \right| \leq \int_0^\pi e^{-r \sin t} \, dt = \int_0^{\pi/2} + \int_{\pi/2}^\pi$

$$= 2 \int_0^{\pi/2} e^{-r \sin t} \, dt.$$

Since $\sin t > t/2$ in $0 \cdots \pi/2$ (Proof?),

$$\left| I_2 \right| \leq 2 \int_0^{\pi/2} e^{-(1/2) r t} \, dt = \frac{4}{r} \left(1 - e^{-r \pi / 4} \right)$$

so that $I_2 \to 0$ as $r \to \infty$.

b) Since

$$I_4 + \pi i = -i \int_0^\pi \left[e^{-\rho \sin t + i\rho \cos t} - 1 \right] dt,$$

$$\left| I_4 + \pi i \right| \leq \int_0^\pi \rho e^\rho \, dt = \rho e^\rho \pi$$

(here we used the estimate $|e^w - 1| \leq |w| e^{|w|}$ from §6, problem 2). If $\rho \to 0$, then $I_4 + \pi i \to 0$, so that $I_4 \to -\pi i$.

7. The length of the path of integration is $\leq 2\pi r$. Thus, by virtue of Theorem 5 in KI, 11, p. 45,

$$\left| \int_{t_r} f(z) \, dz \right| \leq 2\pi r M(r)$$

and the right hand side approaches 0 by hypothesis (for $r \to 0$).

8. The proof is exactly the same as in problem 7.

9. If the amplitude of the radius is α, then $z = te^{i\alpha}$, $t = 0 \cdots 1$. Hence

$$I = e^{i\alpha} \int_0^1 e^{-(1/t)(\cos \alpha - i \sin \alpha)} \, dt$$

$$= e^{i\alpha} \int_0^1 e^{-\cos \alpha/t} \cdot e^{i(\sin \alpha/t)} \, dt.$$

For $\cos \alpha > 0$, i.e. for $-\pi/2 < \alpha < +\pi/2$, the integral converges absolutely. It also converges for $\alpha = \pm\pi/2$, since, for instance,

$$\int_0^1 \sin \frac{1}{t} \, dt = \int_1^{+\infty} \frac{\sin \tau}{\tau^2} \, d\tau$$

and the last integral is evidently convergent. The integral diverges for $\cos \alpha < 0$, i.e. for $+\pi/2 < \alpha < 3\pi/2$. To prove this, disregard the factor $e^{i\alpha}$ and consider the real part of the integral. Setting $-\cos \alpha = c > 0$, $c' = (1 - c^2)^{1/2} \geq 0$, we get

$$\int_0^1 e^{c/t} \cos \left(\frac{c'}{t} \right) \, dt = \int_1^{+\infty} e^{c\tau} \cos (c'\tau) \frac{d\tau}{\tau^2}.$$

At any distance from the origin there are intervals of fixed length $2\pi/3$ in which $\cos (c'\tau) \geq 1/2$. This shows that the last integral cannot converge.

10. The integrals converge if and only if

$$\text{a) } \cos 2\alpha \geq 0, \qquad \text{b) } \cos p\alpha \geq 0.$$

The proof is similar to the one given for problem 9.

11. In this case the definition of the integral as a limit of a sum is not applicable. The integral may be defined as follows. Let $z_1, z_2, \cdots, z_n, \cdots$ be a sequence of points on L different from b and such that $z_n \to b$. The integrals

$$I_n = \int_a^{z_n} f(z) \, dz$$

exist for all n. Assume that for every such sequence (z_n) the sequence (I_n) converges. It is easy to show (cf. §5, problem 8) that $I = \lim I_n$ is independent of the choice of the sequence (z_n). In this case we say that the improper integral $\int_{aL}^b f(z) \, dz$ exists and has the value I.

12. The answer is similar to the one given above. The integral exists and has the value I if for every sequence of points z_n on L such that $z_n \to \infty$, the (proper or improper) integrals

$$I_n = \int_a^{z_n} f(z) \, dz$$

converge to I as $n \to \infty$. If convergence takes place for every such sequence, the number I will be the same for all sequences.

§8. Cauchy's Integral Theorems and Integral Formulas

1. According to Cauchy's theorem the paths of integration may be replaced by the circle $|z - z_0| = r$. A simple computation (reproduced in KI, p. 43) shows that the integral vanishes for $m \neq -1$, equals $2\pi i$ for $m = -1$.

2. Using the decomposition $1/(1 + z^2) = (1/2i) \times [1/(z - i) - 1/(z + i)]$ we obtain without any difficulty the answers:

$$\text{a) } \pi, \qquad \text{b) } -\pi, \qquad \text{c) } 0.$$

3. Choose a point z_0 not on L and a positive number ρ such that $|z_0 - \zeta| > \rho$ for all points ζ on L. For $|z_0 - z| \leq \rho$ we have

$$\frac{f(z) - f(z_0)}{z - z_0} - \frac{3!}{2\pi i} \int_L \frac{\phi(\zeta)}{(\zeta - z_0)^4}\, d\zeta$$

$$= \frac{2!}{2\pi i} \int_L \phi(\zeta) \left\{ \frac{1}{z - z_0} \left(\frac{1}{(\zeta - z)^3} - \frac{1}{(\zeta - z_0)^3} \right) \right.$$

$$\left. - \frac{3}{(\zeta - z_0)^4} \right\} d\zeta$$

Denote this quantity by D. We must show that $D \to 0$ as $z \to z_0$. Set $\zeta - z = A$, $z - z_0 = \eta$. Then $\zeta - z_0 = A + \eta$ and

$$D = \frac{2!}{2\pi i} \int_L \phi(\zeta) \left\{ \frac{(A + \eta)^3 - A^3}{\eta \cdot (A + \eta)^3 \cdot A^3} - \frac{3}{(A + \eta)^4} \right\} d\zeta$$

$$= \frac{2!}{2\pi i} \int_L \phi(\zeta) \frac{(3A^2 + 2A\eta + \eta^2)(A + \eta) - 3A^3}{(A + \eta)^4 \cdot A^3}\, d\zeta.$$

If we compute the numerator, we see that $\eta = z - z_0$ may be taken out before the integral sign, and we get

$$D = (z - z_0) \cdot \frac{2!}{2\pi i} \int_L \phi(\zeta) \, \frac{g(z, z_0, \zeta)}{(\zeta - z)^3 (\zeta - z_0)^4} \, d\zeta$$

where $g(z, z_0, \zeta)$ is an integral rational function of its three arguments. Now let M_1 be an upper bound for $|\phi(\zeta)|$ on L, M_2 an upper bound for $|g(z, z_0, \zeta)|$ for $|z - z_0| \leq \rho$ and ζ on L, d a lower bound for the distance between a point of L, and a point z on the circle $|z - z_0| = \rho$, l the length of L. We have

$$|D| \leq \frac{1}{\pi} \cdot l \cdot M_1 \cdot M_2 \cdot \frac{1}{d^7} \cdot |z - z_0| = k |z - z_0|$$

where k is a fixed constant (depending on z_0 and ρ). It follows that $z \to z_0$ implies $D \to 0$.

4. For $\nu = 1$ the assertion was proved in the preceding answer. Assume it to be true for $\nu - 1$. We must show that

$$\frac{f^{(\nu-1)}(z) - f^{(\nu-1)}(z_0)}{z - z_0}$$

$$= \frac{(\nu + 1)!}{2\pi i} \int_L \frac{\phi(\zeta)}{z - z_0} \left[\frac{1}{(\zeta - z)^{\nu+2}} \right.$$

$$\left. - \frac{1}{(\zeta - z_0)^{\nu+2}} \right] d\zeta$$

$$\to \frac{(\nu + 2)!}{2\pi i} \int_L \frac{\phi(\zeta)}{(\zeta - z)^{\nu+3}} \, d\zeta.$$

As in the preceding proof, we show that the difference of the two integrals may be written in the form

$$(z - z_0) \cdot \frac{(\nu + 1)!}{2\pi i} \int_L \phi(\zeta) \frac{g_1(z, z_0, \zeta)}{(\zeta - z)^{\nu+2}(\zeta - z_0)^{\nu+3}} d.$$

This implies that the absolute value of the difference does not exceed $|z - z_0| \cdot k$; hence it $\to 0$ as $z \to z_0$.

5. Let z_0 be a point of $\mathfrak{G}$. We must show that $f(z)$ is regular at z_0. We take z_0 as the center of a square Q satisfying the conditions mentioned in the theorem. Let $a + bi$ be the left lower vertex of Q, $\zeta = \xi + i\eta$ an arbitrary point of Q. Set

$$F(\zeta) = \int_L f(z) \, dz$$

where the path L consists of a straight segment leading from $a + bi$ to $\xi + bi$ and another straight segment leading from $\xi + bi$ to $\xi + i\eta$. Then

$$F(\zeta) = \int_a^\xi f(x + ib) \, dx + i \int_b^\eta f(\xi + iy) \, dy,$$

so that

$$\frac{1}{i} \frac{\partial F}{\partial \eta} = f(\zeta).$$

By hypothesis we also have

$$F(\zeta) = \int_{L'} f(z) \, dz$$

where the path L' consists of two straight segments, one leading from $a + bi$ to $a + i\eta$, the other from $a + i\eta$ to $\xi + i\eta$. Thus

$$F(\zeta) = \int_b^\eta f(a + iy)\, dy + \int_a^\xi f(x + i\eta)\, dx,$$

so that

$$\frac{\partial F}{\partial \xi} = f(\zeta).$$

Set $F(\zeta) = U(\xi, \eta) + iV(\xi, \eta)$. The relation $\partial F/\partial \xi = -i\ \partial F/\partial \eta = f(\zeta)$ shows that U and V possess continuous partial derivatives satisfying the Cauchy-Riemann differential equations. This means (KI, p. 30) that $F(\zeta)$ is regular in Q and, in particular, at z_0. The same must be true for $f(\zeta) = F'(\zeta)$ (KI, p. 64).

6. By virtue of Cauchy's theorem, any path leading from 0 to 1 and not containing the points $z = \pm i$ may be replaced by a path consisting of the circle $|z - i| = 1$ traversed a certain number of times, the circle $|z + i| = 1$ traversed a certain number of times, and the straight segment leading from 0 to 1. The circles may be traversed either in the positive or in the negative directions; the resulting integrals will be multiples of π (see problem 2). The integral along the segment equals $\pi/4$. Thus our integral may take on any of the values $\pi/4 + k\pi$, $k = 0, \pm 1, \pm 2, \cdots$, and no other values.

7. We have

$$f(z) = \frac{1}{2\pi i} \int_C \frac{d\zeta}{\zeta(\zeta - z)}.$$

Thus $f(0) = 0$. For $0 < |z| < 1$ we have

$$f(z) = \frac{1}{z} \cdot \frac{1}{2\pi i} \cdot \int_C \left[\frac{1}{\zeta - z} - \frac{1}{\zeta}\right] d\zeta = 0.$$

$f(z)$ takes on $|z| = 1$ the boundary values 0, whereas $|\phi(z)| = 1$.

CHAPTER V

EXPANSION IN SERIES

§9. Series with Variable Terms. Uniform Convergence

1. a) Since $| n^z | = n^{\Re(z)}$, the series converges for $\Re(z) > 1$. The series diverges for $\Re(z) < 1$. *Proof*: Assume that $\Re(z_1) = 1 - \delta$, $\delta > 0$, and apply the result of §3, problem 13c, to the series $\Sigma \, 1/n^{z_1}$ and $\Sigma \, 1/n = \Sigma \, 1/n^{z_1} \cdot 1/n^{1-z_1}$. The general term of the series denoted there by $\Sigma \, | \, b_n - b_{n+1} |$ is in our case

$$\left| \frac{1}{n^{1-z_1}} - \frac{1}{(n+1)^{1-z_1}} \right| \leq \frac{1}{n^\delta} \left| \left(1 + \frac{1}{n} \right)^{1-z_1} - 1 \right|.$$

According to §6, problem 3a, we have for $x > 0$

$$e^x > 1 + x, \qquad \text{hence } \log (1 - x) < x,$$

$$\text{hence } \log \left(1 + \frac{1}{n} \right) < \frac{1}{n}$$

so that (cf. §6, problem 2)

$$\left| \left(1 + \frac{1}{n} \right)^{1-z} - 1 \right| = | \, e^{(1-z_1) \, \log \, [1+(1/n)]} - 1 \, |$$

$$\leq \frac{| \, 1 - z_1 \, |}{n} \cdot K,$$

where $K = 2^{|1-z_1|}$. Thus $\Sigma \, | \, b_n - b_{n+1} |$ converges. It follows that the convergence of $\Sigma \, 1/n^{z_1}$ would imply that of $\Sigma \, 1/n$. Our series also diverges for $\Re(z) = 1$, but we can not reproduce the proof here.

b) $f_n(z)/z^n \to 1$ if $| z | < 1$, $f_n(z) \to -1$ if $| z | >$

1. Hence we have convergence for $|z| < 1$, divergence for $|z| > 1$. The series diverges for $|z| = 1$.

c) Answer and proof are the same as for b).

d) For $|z| \neq 1$ the series converges. Along the unit circle there is a dense set of points (consisting of all roots of unity, $z = e^{(p/q)2\pi i}$) at which infinitely many terms are meaningless. At a point $z = e^{2\pi i\gamma}$, γ irrational, the series could converge. (The same remark applies to series e) and f).)

e) For $|z| < 1$, $|f_n(z)|$ is asymptotically equal to $|z^n/n^2|$, for $|z| > 1$ to $1/n^2$. The series converges for $|z| \neq 1$.

f) If $|z| < 1$, $|f_n(z)|$ is asymptotically equal to $|a_n z^n|$. The series converges if and only if the power series $\sum a_n z^n$ converges. If $|z| > 1$, our series converges if and only if $\sum a_n$ converges. In fact, if $|z| > 1$ the convergence of $\sum a_n z^n/(1 - z^n)$ implies that of $\sum a_n/(1 - z^n)$, and hence that of $\sum a_n/(1 - z^n) - \sum a_n z^n/(1 - z^n) = \sum a_n$. On the other hand, if $\sum a_n$ converges so does the series $\sum a_n(1/z)^n/(1 - (1/z)^n)$ (for $|z| > 1$) and the series

$$\sum \frac{a_n z^n}{1 - z^n} = -\sum a_n - \sum \frac{a_n(1/z)^n}{1 - (1/z)^n}.$$

g) It is known that the series converges for a real z. It diverges for $\Im(z) \neq 0$. In fact, setting $z = x + iy$, $y \neq 0$, we have

$$\sin nz = \frac{1}{2i}(e^{inx-ny} - e^{-inx+ny}),$$

$$|\sin nz| \geq \frac{1}{2}(e^{n|y|} - e^{-n|y|}),$$

so that $(\sin nz)/n$ does not approach 0 as $n \to \infty$.

h) The series converges for a real z (since $|\cos nx| \leq 1$), diverges for $\Im(z) \neq 0$. The proof of the last statement is similar to the one given above.

i) The series converges for $z \neq -1, -2, \cdots$. In fact, for such a z and $n > 2|z|$

$$\left| \frac{(-1)^n}{z+n} + \frac{(-1)^{n+1}}{z+n+1} \right| = \left| \frac{1}{(z+n)(z+n+1)} \right|$$

$$\leq \frac{4}{n(n+1)},$$

and $f_n(z) \to 0$ for a fixed z.

k) The series converges for $z \neq -1, -2, \cdots$. In fact,

$$\left| \frac{(-1)^n n}{(z+n)\log n} - \frac{(-1)^{n+1}(n+1)}{(z+n+1)\log(n+1)} \right| \leq \frac{K}{n \log^2 n}$$

(K being a fixed positive number), the series $\sum 1/(n \log^2 n)$ converges and $f_n(z) \to 0$ for a fixed z.

l) The series converges for $|z| < 1$, diverges for $|z| > 1$ since

$$f_1(z) + f_2(z) + \cdots + f_n(z)$$

$$= \frac{z^2 + z^4 + z^8 + \cdots + z^{2^{n+1}-2}}{1 - z^{2^{n+1}}}$$

$$= \frac{z^2 - z^{2^{n+1}}}{(1-z^2)(1-z^{2^{n+1}})}$$

The last expression $\to z^2/(1-z^2)$ if $|z| < 1$, has no limit if $|z| > 1$.

m) The series converges to $2/(z^2 - 1)$ if $|z| > 1$, diverges if $|z| \leq 1$. In fact,

$$\frac{2}{z^2 - 1} - f_1(z) - f_2(z) - \cdots - f_n(z) = \frac{2^{n+1}}{z^{2^{n+1}} - 1}.$$

The last expression $\to 0$ if $|z| > 1$, has no limit for $|z| \leq 1$.

2. Let δ denote an arbitrary positive number. The series converge uniformly in the following (closed) domains:

    a) $\Re(z) \geq 1 + \delta$;

    b), c) $|z| \leq 1 - \delta$;

    d), e) $|z| \leq 1 - \delta$ and $|z| \geq 1 + \delta$;

    f) $|z| \leq 1 - \delta$ and $|z| \geq 1 + \delta$ if $\sum a_n$ converges; $|z| \leq \rho - \delta$ if $\sum a_n$ diverges and ρ denotes the radius of convergence of $\sum a_n z^n$;

    g) $2k\pi + \delta \leq x \leq 2(k + 1)\pi - \delta$, $y = 0$;

    h) the whole real axis;

    i), k) any closed domain not containing any of the points $-1, -2, \cdots$;

    l) $|z| \leq 1 - \delta$;

    m) $|z| \geq 1 + \delta$.

3. Let $\mathfrak{G}_1$ be a bounded closed sub-domain of $\mathfrak{G}$, C a curve situated in $\mathfrak{G}$, containing $\mathfrak{G}_1$ in its interior and having no points in common with $\mathfrak{G}_1$. Let l denote the length of C and d the minimum distance from a point on C to a point in $\mathfrak{G}_1$. For z in $\mathfrak{G}_1$ we have

$$f_\nu'(z) = \frac{1}{2\pi i} \int_C \frac{f_\nu(\zeta)}{(\zeta - z)^2} \, d\zeta,$$

and hence

$$| f'_{n+1}(z) + \cdots + f'_{n+k}(z) |$$

$$\leq \frac{1}{2\pi} \int_C \frac{| f_{n+1}(\zeta) + \cdots + | f_{n+k}(\zeta) |}{| \zeta - z |^2} | d\zeta |.$$

Let there be given a $\epsilon > 0$. By hypothesis there exists an n_0 such that the numerator of the fraction in the integral $< \epsilon$ for ζ on C, $n > n_0$ and $k \geq 1$. Hence

$$| f'_{n+1}(z) + \cdots + f'_{n+k}(z) | \leq \frac{l}{2\pi d^2} \epsilon$$

for $n > n_0$, $k \geq 1$ and z in $\mathfrak{G}_1$ This proves the assertion.

4. The proof follows word for word the one given for §3, problem 12.

5. Since $| s_n(z) | \leq K$ and $h_{n+1} = a_{n+1} \to 0$, $s_n \cdot h_{n+1} \to 0$ uniformly in G. Since $| s_n(z)[h_{n+1}(z) - h_n(z)] | \leq K(a_n - a_{n+1})$ the series $\sum s_n(z)[h_{n+1}(z) - h_n(z)]$ converges uniformly in $\mathfrak{G}$.

6. We make use of the theorem of problem 4. Set

$$f_n(z) = \frac{(-1)^n}{n^\epsilon}, \qquad h_n(z) = \frac{1}{n^{\epsilon+z}}.$$

The series $\sum f_n(z)$ converges. Its partial sums must be bounded (uniformly in the whole plane since $f_n(z)$ is constant). Set $\delta = 2\epsilon$. If $\Re(z) = \rho \geq 0$, we have that $| h_n(z) | \leq 1/n^{\epsilon+\rho}$, so that (cf. problem 1a) $s_n \cdot h_{n+1} \to 0$ uniformly. If z also satisfies an inequality $| z | \leq R$, then

$$| h_n(z) - h_{n+1}(z) | = \left| \frac{1}{n^{\epsilon+z}} - \frac{1}{(n+1)^{\epsilon+z}} \right|$$

$$\leq \frac{1}{n^{\epsilon+\rho}} \left| \left(1 + \frac{1}{n} \right)^{\epsilon+z} - 1 \right| \leq \frac{K}{n^{1+\epsilon+\rho}}.$$

It follows that the series $\sum s_n(z)[h_n(z) - h_{n+1}(z)]$ converges uniformly. Therefore

$$\sum f_n(z) h_n(z) = \sum \frac{(-1)^n}{n^{\delta+z}}$$

converges uniformly for $\Re(z) \geq 0$, $|z| \leq R$. This means that $\sum (-1)^n/n^z$ converges uniformly for $\Re(z) \geq 0$, $|z| \leq R$.

7. The proof is very similar to the one given above. Set

$$f_n(z) = \frac{a}{n^{z_0}}, \qquad h_n(z) = \frac{1}{n^{z-z_0}}.$$

8. a) The disjoint convergence domains, $|z| < 1$ and $|z| > 1$, were found in problem 1c. The proof of the fact that both functions represented by the series possess the natural boundary $|z| = 1$ requires deeper theorems.

b) The series represents the function 1 in $|z| < 1$, the function 0 in $|z| > 1$.

c) The series represents the function $f_1(z)$ in $|z| < 1$, $f_2(z)$ in $1 < |z| < 1 + \delta$.

§10. Expansion in Power Series

1. These problems are of the following form: the expansions

$$\omega = \phi(z) = a_0 + a_1 z + a_2 z^2 + \cdots$$

$$w = g(\omega) = b_0 + b_1 \omega + b_2 \omega^2 + \cdots$$

are known, the expansion $w = g(\phi(z)) = c_0 + c_1 z + c_2 z^2 + \cdots$ has to be determined. This can be done either by computing $(\phi(z))^2 = \omega^2$, $(\phi(z))^3 = \omega^3$, $\cdots$ and substituting these values in $g(w)$, or by computing the derivatives

$$w' = g'(\omega) \cdot \phi'(z), \ w'' = g''(\omega) \cdot (\phi'(z))^2 + g'(\omega) \cdot \phi''(z),$$

$$w''' = \cdots$$

at $z = 0$. In many special cases, however, various simplifications will be found.

a) $e^{z/(1-z)} = e^{z+z^2+\cdots} = e^z \cdot e^{z^2} \cdot \ \cdots$

$$= 1 + z + \frac{3}{2} z^2 + \frac{13}{6} z^3 + \frac{73}{24} z^4$$

$$+ \frac{501}{120} z^5 + \cdots$$

or

$$= \sum_{n=1}^{\infty} \frac{1}{k!} \left(\frac{z}{1-z} \right)^k$$

$$= 1 + \sum_{n=1}^{\infty} \left(\sum_{\nu=1}^{n} \binom{n-1}{\nu-1} \frac{1}{\nu!} \right) z^n.$$

b) $\sin \dfrac{1}{1-z} = \sin \left(1 + \dfrac{z}{1-z}\right)$

$\qquad = \sin 1 \cdot \cos \dfrac{z}{1-z} + \cos 1 \cdot \sin \dfrac{z}{1-z};$

$\cos \dfrac{z}{1-z} = 1 - \dfrac{1}{2!}\left(\dfrac{z}{1-z}\right)^2 + \dfrac{1}{4!}\left(\dfrac{z}{1-z}\right)^4$

$\qquad\qquad\qquad\qquad\qquad\qquad - + \cdots$

$\qquad = 1 - \dfrac{z^2}{2} - z^3 - \dfrac{35}{24}z^4 - \dfrac{11}{6}z^5 - \cdots ;$

$\sin \dfrac{z}{1-z} = \left(\dfrac{z}{1-z}\right) - \dfrac{1}{3!}\left(\dfrac{z}{1-z}\right)^3 + \cdots$

$\qquad = z + z^2 + \dfrac{5}{6}z^3 + \dfrac{1}{2}z^4 + \dfrac{1}{120}z^5 - \cdots .$

Setting $\sin 1 = \sigma$, $\cos 1 = \gamma$ we have

$\sin \dfrac{1}{1-z} = \sigma + \gamma z + \left(\gamma - \dfrac{1}{2}\sigma\right)z^2 + \left(\dfrac{5}{6}\gamma - \sigma\right)z^3$

$\qquad + \left(\dfrac{1}{2}\gamma - \dfrac{35}{24}\sigma\right)z^4 + \left(\dfrac{1}{120}\gamma - \dfrac{11}{6}\sigma\right)z^5 + \cdots$

c) $e^{(e^z)} = e \cdot e^z \cdot e^{(1/2)z^2} \cdot e^{(1/6)z^3} \cdot \cdots$

$\qquad = e\left[1 + z + \dfrac{z^2}{2!} + \cdots\right]\left[1 + \dfrac{z^2}{2} + \dfrac{z^4}{8} + \cdots\right]$

$$\cdot \left[1 + \frac{z^3}{6} + \cdots \right] \cdots$$

$$= e \left[1 + z + z^2 + \frac{5}{6} z^3 + \frac{5}{8} z^4 + \frac{13}{30} z^5 + \cdots \right]$$

(This result could be obtained faster by direct differentiation.)

d) The successive derivatives of $\log (1 + e^z)$ are

$$\frac{e^z}{1 + e^z}; \quad \frac{e^z}{(1 + e^z)^2}; \quad \frac{e^z - e^{2z}}{(1 + e^z)^3}$$

$$\frac{e^z - 4e^{2z} + e^{3z}}{(1 + e^z)^4}; \quad \frac{e^z - 11e^{2z} + 11e^{3z} - e^{4z}}{(1 + e^z)^5}; \cdots$$

so that

$$\log (1 + e^z) = \log 2 + \frac{z}{2} + \frac{z^2}{8} - \frac{z^4}{192} + \cdots .$$

(The only odd power in this expansion is $z/2$, since $\log (1 + e^z) - z/2$ is an even function.)

e)
$$(\cos z)^{1/2} = \left[1 - \left(\frac{z^2}{2} - \frac{z^4}{24} + \cdots \right) \right]^{1/2}$$

$$= 1 - \frac{z^2}{4} - \frac{z^4}{96} + \cdots$$

f)
$$e^{z \sin z} = 1 + z \sin z + \frac{1}{2} z^2 \sin^2 z + \cdots$$

$$= 1 + z^2 + \frac{z^4}{3} + \cdots .$$

2. a) Let log* $2a$ denote a definite logarithm of $2a$. There exists exactly one function $f(z)$ which is regular at $z = 0$, coincides in a neighborhood of $z = 0$ with one of the branches of $\log [a + (a^2 + z^2)^{1/2}]$, and is such that $f(0) = \log^* 2$. For this function

$$f'(z) = \frac{z}{[a + (a^2 + z^2)^{1/2}](a^2 + z^2)^{1/2}} = \frac{z[(a^2 + z^2)^{1/2} - a]}{z^2(a^2 + z^2)^{1/2}}$$

$$= \frac{1}{z} - \frac{1}{z[1 + (z/a)^2]^{1/2}} = - \sum_{n=1}^{\infty} \binom{-1/2}{n} \frac{z^{2n-1}}{a^{2n}}.$$

Hence

$$f(z) = \log^* 2 + \sum_{n=1}^{\infty} \frac{1}{2n} \binom{-1/2}{n} \left(\frac{z}{a}\right)^{2n}.$$

b) $\dfrac{1}{2}\left(z + \dfrac{z^2}{2} + \dfrac{z^3}{3} + \cdots\right)^2 = \dfrac{1}{2} \displaystyle\sum_{n=2}^{\infty} \left(\dfrac{1}{1 \cdot (n - 1)} + \cdots\right.$

$$+ \frac{1}{k(n - k)} + \cdots + \frac{1}{(n - 1) \cdot 1}\bigg) z^n$$

$$= \sum_{n=2}^{\infty} \frac{1}{n}\left(1 + \frac{1}{2} + \cdots + \frac{1}{n - 1}\right) z^n.$$

c) In a similar way one obtains

$$(\arctan z)^2 = \sum_{n=1}^{\infty} \frac{(-1)^{n-1}}{n}\left(1 + \frac{1}{3} + \cdots + \frac{1}{2n - 1}\right) z^{2n},$$

and

d) $\arctan z \cdot \log(1 + z^2)$

$$= 2 \sum_{n=1}^{\infty} \frac{(-1)^{n-1}}{2n + 1} \left(1 + \frac{1}{2} + \frac{1}{3} + \cdots + \frac{1}{2n}\right) z^{2n+1}.$$

e) The simplest way to obtain the expansion of $\cos^2 z$ is as follows:

$$2 \cos^2 z = 1 + \cos 2z = 1 + \sum_{n=0}^{\infty} (-1)^n \frac{2^{2n}}{(2n)!} z^{2n}.$$

The expansion of $\sin^2 z$ can be obtained from the identity $\sin^2 z = (1 - \cos 2z)/2$. It is simpler to set

$$\sin^2 z = 1 - \cos^2 z = \frac{1}{2} \sum_{n=1}^{\infty} (-1)^{n-1} \frac{2^{2n}}{(2n)!} z^{2n}.$$

f) The expansion

$$\frac{1}{\cos z} = \sum_{n=0}^{\infty} (-1)^n \frac{E_{2n}}{(2n)!} z^{2n}$$

defines the *Euler numbers* E_{2n}. The identity

$$\left(1 - \frac{z^2}{2!} + \frac{z^4}{4!} - + \cdots\right)\left(E_0 - \frac{E_2}{2!} z^2 \right.$$

$$\left. + \frac{E_4}{4!} z^4 - + \cdots\right) \equiv 1$$

shows that $E_0 = 1$ whereas the other Euler numbers (integers) can be computed from the recursion formulas

$$E_0 + \binom{2n}{2} E_2 + \binom{2n}{4} E_4 + \cdots + \binom{2n}{2n} E_{2n} = 0.$$

Thus

$$E_2 = -1, \quad E_4 = 5, \quad E_6 = -61, \quad E_8 = 1385, \cdots.$$

g) The expansion

$$\frac{z}{e^z - 1} = \frac{1}{1 + z/(2!) + z^2/(3!) + \cdots} = \sum_{n=0}^{\infty} \frac{B_n}{n!} z^n.$$

defines the *Bernoulli numbers* B_n which satisfy the recursion formulas

$$\binom{n}{0}B_0 + \binom{n}{1}B_1 + \cdots + \binom{n}{n-1}B_{n-1} = 0$$

(for $n \geq 2$), whereas $B_0 = 1$. Thus

$$B_1 = -\frac{1}{2}, \quad B_2 = \frac{1}{6}, \quad B_3 = 0, \quad B_4 = -\frac{1}{30}, \cdots.$$

h) Since $\tan z = \cot z - 2 \cot 2z$, $z \cot z = iz + 2iz/(e^{2iz} - 1)$, we can obtain the expansion of $\tan z$ from g). Since $\log \cos z = -\int_0^z \tan z \, dz$ we get

$$\log \cos z = \sum_{n=1}^{\infty} (-1)^n \frac{2^{2n}(2^{2n} - 1)B_{2n}}{2n(2n)!} z^{2n}.$$

i) Using g) and the identity $e^z/(e^z + 1) = (1/z)[2z/(e^{2z} - 1)] - (1/z)[z/(e^z - 1)] + 1$ we obtain

$$\frac{e^z}{e^z + 1} = 1 + \sum_{n=1}^{\infty} \frac{(2^n - 1)B_n}{n!} z^{n-1}$$

$$= \frac{1}{2} + \sum_{n=1}^{\infty} \frac{(2^{2n} - 1)B_{2n}}{(2n)!} z^{2n-1}.$$

k) Using g), h) and the identity $1/(\sin z) = \cot z + \tan (z/2)$ we obtain

$$\frac{z}{\sin z} = \sum_{n=0}^{\infty} (-1)^{n-1} \frac{2(2^{2n-1} - 1)B_{2n}}{(2n)!} z^{2n}.$$

3. a) The series $f(z) = \sum_{n=1}^{\infty} 1/n^z$ converges uniformly in the half-plane $\Re(z) \geq 1 + \delta$ $(\delta > 0)$. The same is true for all series obtained by formal differentiation. Hence

$$f(z) = \sum_{k=0}^{\infty} a_k(z - 2)^k, \qquad a_k = \frac{(-1)^k}{k!} \sum_{n=1}^{\infty} \frac{(\log n)^k}{n^2}.$$

$(a_0 = \pi^2/6$; the other coefficients can not be expressed in a simpler form).

b) By Weierstrass' double-series theorem (KI, 20, p. 83) the identity

$$\sum_{n=1}^{\infty} b_n \frac{z^n}{1 - z^n} = \sum_{k=1}^{\infty} a_k z^k$$

implies that $a_k = \sum b_d$, where d runs over all divisors of k, including 1 and k. (Example: $a_6 = b_1 + b_2 + b_3 + b_6$). In particular, if $b_n \equiv 1$, $a_k = \tau(k) =$ number of divisors of k. Thus

$$\sum_{n=1}^{\infty} \frac{z^n}{1 - z^n} = z + 2z^2 + 2z^3 + 3z^4 + 2z^5 + 4z^6$$

$$+ 2z^7 + 4z^8 + 3z^9 + \cdots .$$

c) According to b) $\sum_{n=1}^{\infty} \phi(n) \ z^n/(1 - z^n) = \sum_{k=1}^{\infty} a_k z^k$ where $a_k = \sum_{d|k} \phi(d)$. ($d \mid k$ denotes that d is a divisor of k.) It is known that $\sum_{d|k} \phi(d) = k$. Hence

$$\sum_{n=1}^{\infty} \phi(n) \frac{z^n}{1 - z^n} = z + 2z^2 + 3z^3 + \cdots = \frac{z}{(1 - z)^2}.$$

4. Assume that C encloses infinitely many zeros of $f(z) - a$. Then these zeros possess a limit point z_0 which lies either on C or in the domain interior to C. Since z_0 is an interior point of $\mathfrak{G}$, the identity theorem for analytic functions (KI, 21) implies that $f(z) \equiv a$ contrary to the hypothesis.

5. a) No. The function would have to coincide with $w \equiv 0$ (by virtue of the identity theorem for analytic functions, KI, 21) and could not assume the value 1.

b) No. The function would have to coincide with $w \equiv 0$ and could not assume the values $1/2$, $1/4$, $\cdots$.

c) No, since the function would have to coincide with $w \equiv z$ and would have to differ from this function at $z = 1/3$, $1/5$, $\cdots$.

d) Yes. In fact, $f(z) = 1/(z + 1)$.

6. Since $f(z) = a_\alpha (z - z_0)^\alpha + \cdots$, $a_\alpha \neq 0$,

$$\int_{z_0}^{z} f(z)\, dz = \frac{a_\alpha}{\alpha + 1} (z - z_0)^{\alpha + 1} + \cdots,$$

$$\int_{z_1}^{z} f(z)\, dz = \int_{z_1}^{z_0} f(z)\, dz + \int_{z_0}^{z} f(z)\, dz =$$

$$c + \frac{a_\alpha}{\alpha + 1} (z - z_0)^{\alpha + 1} + \cdots.$$

Thus z_0 is a zero of order $(\alpha + 1)$ of $F_0(z)$ and a c-point of order $(\alpha + 1)$ of $F_1(z)$, where $c = \int_{z_1}^{z_0} f(z)\, dz$.

7. Because $z^{1/2}$ is not regular at $z = 0$. (See the definition in KI, 21.)

8. There exist several theorems yielding lower bounds of increasing accuracy. A first result may be obtained as follows.

If $f(0) = 0$, i.e.

$$f(z) = a_\alpha z^\alpha + \cdots, \ a^\alpha \neq 0, \ \alpha \geq 1,$$

then $f(z)$ and the function

$$f_1(z) = a_\alpha + a_{\alpha+1}z + \cdots$$

possess the same zeros, except at $z = 0$. Thus we lose no generality in assuming that $a_0 \neq 0$.

Let ρ be a number such that $0 < \rho < r$ and M the maximum of $|f(z)|$ on $|z| = \rho$. We shall show that $f(z) \neq 0$ for

$$|z| < \frac{|a_0|}{|a_0| + M} \rho.$$

By Cauchy's inequality (KI, p. 77) $|a_n| \leq M/\rho^n$. Hence we have for all values of z satisfying the preceding condition

$$|f(z) - a_0| \leqq |a_1||z| + \cdots \leq M\left[\frac{|z|}{\rho} + \cdots\right]$$

$$= M \cdot \frac{|z|}{\rho - |z|}$$

so that

$$|f(z) - a_0| < M \ \frac{\dfrac{|a_0|}{|a_0| + M}\rho}{\rho - \dfrac{|a_0|\rho}{|a_0| + M}} = |a_0|$$

and

$$|f(z)| = |a_0 - (f(z) - a_0)|$$
$$\geq |a_0| - |f(z) - a_0| > 0.$$

Considerably stronger results will be found in the paper by E. Landau, Über eine Aufgabe aus der Funktionentheorie, *The Tohoku Mathematical Journal*, vol. 5, 1914, pp. 97-116.

§11. Behaviour of Power Series on the Circle of Convergence

1. a) $\sum z^n/n^2$, b) $\sum z^n$, c) $\sum z^n/n$ (The exceptional point is $z = 1$, see also §4, problem 5.)

2. All four series have the radius of convergence 1. The series converge at all points of the circle of convergence except at a) $z = -1$, b) and c) $z = \pm i$, d) $z = +1$. *Proof*. Use the result of §4, problem 8, after setting a) $z' = z$, b) and c) $z' = -z^2$.

3. Set $r/2 = z_0$,

$$f(z) = \sum_{n=0}^{\infty} a_n z^n = \sum_{k=0}^{\infty} b_k (z - z_0)^k.$$

By Taylor's theorem

$$b_k = \sum_{n=0}^{\infty} \binom{n+k}{k} a_{n+k} z_0^n .$$

Assume that $z_1 = +r$ is a regular point of $f(z)$. Then the radius of convergence of $\sum b_k (z - z_0)^k$ exceeds $r/2$, so that this series converges for $z - z_0 = z_0 + \alpha$, α being a sufficiently small positive number. All terms of the series

$$\sum_k b_k (z_0 + \alpha)^k = \sum_k \left[\sum_n \binom{n+k}{k} a_{n+k} z_0^n \right] (z_0 + \alpha)^k$$

are real and non-negative, so that the terms of the series may be rearranged without destroying convergence. Arranging the terms in ascending order of $n + k = m$, we obtain the series

$$\sum_{m=0}^{\infty} a_m \left[\sum_{k=0}^{m} \binom{m}{k} z_0^{m-k} (z_0 + \alpha)^k \right] = \sum_{m=0}^{\infty} a_m (r + \alpha)^m.$$

This series, however, cannot converge, since $r + \alpha$ lies outside the circle of convergence of $\sum a_n z^n$. Thus we arrive at a contradiction.

4. The proof given in KI (p. 101) for a) can be adopted to the other cases. Another proof uses the result of the preceding problem. It shows that $z = 1$ is a singular point of $f(z) = \sum z^{n!}$. Let z_0 be a q-th root of unity, $z_0 = e^{2\pi i (p/q)}$. Then $f_0(z) = f(z/z_0)$ is singular at z_0. So is $f(z)$, for $f(z) - f_0(z)$ is a polynomial. Since every point on the unit circle is a limit point of roots of unity, all points on the unit circle are singular. The same argument holds for b) and c), if we restrict q to the values 2^n and $g_1 \cdot g_2 \cdots g_n$, respectively. To prove the assertion concerning d), differentiate this series twice. The resulting series is identical with b). (Note that d) converges absolutely at all points of the unit circle.)

5. a) It will suffice to show that $|f(z)| \to \infty$ as z approaches any root of unity $z_0 = e^{2\pi i (p/q)} (q > 0, p$ and q relatively prime) along a radius. (See the proof in KI, 24, p. 101.) We set $z = \rho z_0$, $0 \leq \rho < 1$, and divide the series in two series, T_1 and T_2, the first containing those terms for which n is divisible by q. Then

$$T_1 = \sum_\nu \frac{z^{q\nu}}{1 - z^{q\nu}} = \sum \frac{\rho^{q\nu}}{1 - \rho^{q\nu}}$$

so that

$$\lim_{\rho \to 1} (1 - \rho)T_1 = \lim_{\rho \to 1} \frac{1 - \rho}{1 - \rho^q} (1 - \rho^q)T_1$$

$$= \frac{1}{q} \lim_{y \to 1} (1 - y) \sum_\nu \frac{y^\nu}{1 - y^\nu}.$$

Since, for $0 < y < 1$,

$$(1 - y) \sum_\nu \frac{y^\nu}{1 - y^\nu}$$

$$= \sum_\nu \frac{y^\nu}{1 + y + \cdots + y^{\nu-1}} > \sum \frac{y^\nu}{\nu} = \log \frac{1}{1 - y}.$$

we have $(1 - \rho)T_1 \to \infty$ as $\rho \to +1$.

If n is not divisible by q, then there exists a positive α such that $|1 - z^n| > \alpha$ (for $z = \rho z_0$). This follows from the fact that z^n lies on one of the $q - 1$ radii which lead to the points $e^{2\pi i(\nu/q)}$, $\nu = 1, 2, \cdots, q - 1$. (If $q > 2$ we may set $\alpha = \sin(2\pi/q)$.) Thus

$$|(1 - \rho)T_2| < (1 - \rho) \sum \frac{\rho^n}{\alpha} \leqq \frac{1}{\alpha}.$$

It follows that $|(1 - \rho)(T_1 + T_2)| = \{(1 - \rho) |f(z)|\} \to +\infty$ as $\rho \to +1$. Hence every root of unity is a singular point of $f(z)$, and the unit circle is a natural boundary.

b) The proof is more complicated. It will be found in the paper by K. Knopp, Über Lambertsche

Reihen, *Journal f. d. reine und angew. Math.*, vol. 142, 1913, p. 291.

6. For $|z| = 1$ the absolute value of the nth term is $1/n$, so that the series fails to converge absolutely.

For $z = 1$ we have the series

$$- 1 - \frac{1}{2} - \frac{1}{3} + \frac{1}{4} + \frac{1}{5} + \cdots + \frac{1}{8} - \frac{1}{9} - \cdots$$

$$- \frac{1}{15} + \frac{1}{16} + \cdots + \frac{1}{24} - \cdots .$$

Terms with the denominators k^2, $(k^2 + 1)$, $\cdots$, $(k + 1)^2 - 1$ have the same sign as $(-1)^k$ ($k = 1, 2, \cdots$). Denote the sum of these terms by $(-1)^k g_k$. It suffices to show that $\sum (-1)^k g_k$ converges (Why?). The convergence of this series, however, follows from the fact that

$$0 < g_k < \frac{(k + 1)^2 - k^2}{k^2} = \frac{2k + 1}{k^2} \to 0 \text{ as } k \to + \infty$$

and

$$g_k - g_{k+1} = \left(\frac{1}{k^2} - \frac{1}{(k + 1)^2}\right) + \cdots + \left(\frac{1}{(k + 1)^2 - 1}\right.$$

$$- \frac{1}{(k + 2)^2 - 3}\right) - \frac{1}{(k + 2)^2 - 2} - \frac{1}{(k + 2)^2 - 1}$$

$$> (2k + 1)\left(\frac{1}{(k + 1)^2} - \frac{1}{(k + 2)^2 - 2}\right)$$

$$- \frac{1}{(k + 2)^2 - 2} - \frac{1}{(k + 2)^2 - 1} > 0.$$

For $z \neq 1$, $|z| = 1$ we use the result of §3, problem 14, setting $a_n = (-1)^{[n^{1/2}]} z^n$, $b_n = 1/n$. We have

$$s_n = -z - z^2 - z^3 + z^4 + \cdots + (-1)^{[n^{1/2}]} z^n.$$

Let p denote the largest integer for which $(p + 1)^2 \leq n + 1$. Then

$$s_n = -z \frac{1 - z^3}{1 - z} + z^4 \frac{1 - z^5}{1 - z} + z^9 \frac{1 - z^7}{1 - z} + - \cdots$$

$$\pm z^{p^2} \frac{1 - z^{2p+1}}{1 - z} \mp (z^{(p+1)^2} + \cdots + z^n),$$

$$|s_n| < \frac{2p}{|1 - z|} + 2p + 3 < K \cdot p < K\, n^{1/2}$$

so that the sequence $(s_n/n^{1/2})$ is bounded. Furthermore, $n^{1/2} b_n = 1/n^{1/2} \to 0$ and $\sum n^{1/2}(b_n - b_{n+1}) = \sum 1/[n^{1/2}(n + 1)]$ converges. It follows that the series $\sum a_n b_n$ converges.

7. Let G be a given positive number. There exists an m such that $b_0 + b_1 + \cdots + b_m > G + 1$. The polynomial $b_0 + b_1 x + \cdots + b_m x^m$ exceeds the value $G + 1$ at $x = 1$; hence there exists a positive $x_0 < 1$ such that $b_0 + b_1 x + \cdots + b_m x^m > G$ for $x \geq x_0$. It follows that $h(x) > G$ for $x \geq x_0$.

8. Set $a_{n+1} + a_{n+2} + \cdots = r_n$. Using summation by parts, we have

$$T = \sum_{\nu=n+1}^{n+p} a_\nu z^\nu = \sum_{\nu=n+1}^{n+p} (r_{\nu-1} - r_\nu) z^\nu$$

$$= - \sum_{\nu=n+1}^{n+p} r_\nu(z^\nu - z^{\nu+1}) + r_n z^{n+1} - r_{n+p} z^{n+p+1}.$$

For a given $\epsilon > 0$ we can find an n_0 such that $|r_n| < \epsilon/3$ for $n \geq n_0$. For $z = x$, $0 \leq x \leq 1$ and $n \geq n_0$ we have

$$|T| \leq \frac{\epsilon}{3}(x^{n+1} - x^{n+p+1}) + \frac{\epsilon}{3} + \frac{\epsilon}{3} < \epsilon$$

which proves the assertion.

9. Using the same notations as above, we have

$$|T| \leq \frac{2}{3}\epsilon + \frac{\epsilon}{3}\sum_{\nu=n+1}^{n+p}|z^\nu - z^{\nu+1}| \leqq \frac{2}{3}\epsilon + \frac{\epsilon}{3}\cdot K$$

where K is the constant from §1, problem 13. This inequality implies the assertion.

10. Let $\epsilon > 0$ be given. By virtue of the result of problem 9 there exists an m such that

$$|r_m(z)| = |a_{m+1}z^{m+1} + a_{m+2}z^{m+2} + \cdots| < \frac{\epsilon}{2}$$

for all z within the triangle. After choosing m we may choose a $\delta > 0$ such that

$$|s_m(z) - s_m| = |(a_0 + a_1z + \cdots + a_mz^m)$$

$$- (a_0 + a_1 + \cdots + a_m)| < \frac{\epsilon}{2}$$

for $|z - 1| < \delta$. Thus we have

$$\left|\sum a_n z^n - \sum a_n\right| < \epsilon$$

provided z lies within the triangle and $|z - 1| < \delta$.

CHAPTER VI

CONFORMAL MAPPING

§12. Linear Functions. Stereographic Projection

1. a) $w + 5i/2 = 3(z + 5i/2)$, b) $w + (3/5)(1 - 2i)$
$= (i/2)[z + (3/5)(1 - 2i)]$, c) $w - b/(1 - a) =$
$a[z - b/(1 - a)]$.

2. $w = (1 - i)z - 1$. The function is uniquely determined. It must be of the form $w = az + b$ and for $z = 0, 1$ it must take on the values $-1, -i$.

3. Proof. A similarity transformation is determined by the (distinct) images, w_1 and w_2 , of two distinct points, z_1 and z_2 . The conditions

$$w_1 = az_1 + b, \qquad w_2 = az_2 + b$$

yields

$$a = \frac{w_2 - w_1}{z_2 - z_1}, \qquad b = \frac{w_2 z_1 - w_1 z_2}{z_2 - z_1}.$$

The transformation, $w = az + b$ may be written in the form

$$\begin{vmatrix} w & z & 1 \\ w_1 & z_1 & 1 \\ w_2 & z_2 & 1 \end{vmatrix} = 0.$$

4. a) Straight line: $\Re(z) = 1/2$,
   b) Circle: $|z - 8/3| = 4/3$,
   c) Circle: $|z| = 1/r$,
   d) Straight line: $\Re(z\bar{z}_0) = 1$,

e) This circle is taken into itself,

f) Circle: $\alpha + \beta x + \gamma y + \alpha'(x^2 + y^2)$. (If $\alpha' = 0$, we have a straight line in the w plane, if $\alpha = 0$, a straight line in the z plane. In the theory of linear transformations straight lines are considered to be special cases of circles.)

g) Circle: $|\, 2\alpha z - 1\,| = 1$ if $\alpha \neq 0$. If $\alpha = 0$, the straight line is taken into itself.

h) Family of circles: $|\, \lambda(1 + i)z - 1\,| = 1$, λ real and $\neq 0$. These circles pass through the origin and are tangent to the straight line $y = x$.

i) Family of circles: $|\, z/(\lambda i z_0) - 1\,| = 1$, λ real and $\neq 0$.

k) Family of all circles through 0 and $1/\bar{z}_0$.

l) Family of circles through $1/\bar{z}_0$. (If $z_0 = 0$, all straight lines.)

m) Family of circles through $1/\bar{z}_0$ and $1/\bar{z}_1$.

n) One of the 8 circular arc triangles formed by the circles passing through 0 and two of the three points $1/\bar{z}_1$, $1/\bar{z}_2$, $1/\bar{z}_3$.

o) Cissoid: $y^2 = (2px^3)/(1 - 2px)$,

p) Lemniscate: $(x^2 + y^2)^2 - (x^2 - y^2) = 0$,

q) Cissoid: $y(x^2 + y^2) - x^2 = 0$.

5. The answers are similar to the ones given to problem 4. (The curves obtained in 0), p), q) have no special names.)

6. a) $+1$, $+i$, -1, $-i$ are taken into points on the equator (of longitudes: $0°$, $90°$, $180°$, $270°$). The image of $z = x + iy$ has latitude $\beta = 2$ arc tan $|\, z\,| - \pi/2$, longitude $\lambda = $ am z, $(0 \leq $ arc tan $|\, z\,| < \pi/2)$.

b) Southern hemisphere, equator, northern hemisphere.

c) $-90° < λ < +90°$ ("right" hemisphere), $λ = ±90°$, "left" hemisphere.

d) Eastern hemisphere, meridians $λ = 0°$, $λ = 90°$, western hemisphere.

e) Latitude circle.

f) Meridian.

7. a) Diametrically opposite points on a latitude circle.

b) Points with the same latitude and opposite longitudes.

c) Points with the same longitude and opposite latitudes.

d) Points situated symmetrically with respect to the diameter of the equator joining the points $λ = 0$ and $λ = 180°$.

8. a) A pencil of circles through the north pole possessing there a common tangent.

b) A reflection with respect to the plane of the prime meridian.

c) A reflection with respect to the plane of the meridians $λ = ±90°$.

d) A reflection with respect to the equatorial plane.

e) A spherical triangle whose three sides meet at the north pole.

9. a) The ray am $z = λ$, the circle $|z| = \tan(β/2 + π/4)$.

b) The pair of points $z, -1/\bar{z}$.

c) Circles intersecting the unit circle in two diametrically opposite points.

d) A circular arc triangle, formed of arcs of circles satisfying the condition mentioned in c).

e) Let A and B be the points at which the image k' of k intersects the straight line through the origin and the center of k'. There exists a uniquely determined pair of points z_0, z_0' harmonic to A, B and such that $z_0 = -1/\bar{z}_0'$. If z_0 lies within k', then z_0 is the image of M_0.

f) The pencil of circles passing through the image z_0 of P and through $-1/\bar{z}_0$. (The pencil of straight lines through the origin if P is the North Pole or the South Pole.)

g) The point $z = \tan(\beta/2 + \pi/4) \cdot (\cos \lambda + i \sin \lambda)$.

10. a) We use a Cartesian coordinate system (ξ, η, ζ) with origin at the center of the sphere. Let the ξ-axis (η-axis) be parallel to the x-axis (y-axis) in the z-plane. The image of $z = x + iy$ has the coordinates

$$\xi = \frac{x}{[4(x^2 + y^2) + 1]^{1/2}}, \qquad \eta = \frac{y}{[4(x^2 + y^2) + 1]^{1/2}},$$

$$\zeta = -\frac{1}{2[4(x^2 + y^2) + 1]^{1/2}}$$

b) and c) only at $z = 0$. In fact
$$d\xi^2 + d\eta^2 + d\zeta^2 = [(4y^2 + 1)dx^2 - 8xy\,dx\,dy + (4x^2 + 1)dy^2] \cdot (4x^2 + 4y^2 + 1)^2.$$

The right hand side has the form $\phi(x, y)(dx^2 + dy^2)$ only if $x = y = 0$.

11. $(w - \zeta_1)/(w - \zeta_2) = a(z - \zeta_1)/(z - \zeta_2)$, $a \neq 0$. If one of the fixed points is ∞, the corresponding differences are replaced by 1. The transformation is called elliptic if $|a| = 1$, hyperbolic if a is real and

positive, loxodromic if it is neither elliptic nor hyperbolic (hence, a combination of both). (The proof will be found in C, Chapter I.)

12. The transformation is uniquely determined and has the form

$$\frac{w - w_1}{w - w_2} \div \frac{w_3 - w_1}{w_3 - w_2} = \frac{z - z_1}{z - z_2} \div \frac{z_3 - z_1}{z_3 - z_2}.$$

If one of the given points is ∞, the corresponding differences are replaced by 1. It is clear that this mapping is a linear transformation satisfying our conditions. That it is the only such function follows from the invariance of the cross-ratio (see C, §10–12).

13. z_2 is the reflection of z_1 with respect to the circle k if and only if every circle containing z_1 and z_2 is orthogonal to k. The assertion follows from the fact that a linear transformation is conformal and takes circles into circles.

14. The transformation must take $z = \infty$ into $w = -1$ (since $z = \infty$ is the reflection of $z = 0$ with respect to $|z| = 1$, see problem 13). Thus the points $z = 0$, 1, ∞ must be taken into $w = 1/2$, 0, -1 respectively. The result of problem 12 yields the uniquely determined transformation $w = -(z - 1)/(z + 2)$.

15. $w = (\alpha z + \beta)/(\gamma z + \delta)$, α, β, γ, δ real, $\alpha\delta - \beta\gamma > 0$. In fact this transformation takes the real axis into itself and $z = i$ into $w = (\alpha i + \beta)/(\gamma i + \delta)$. Since $\Im\{(\alpha i + \beta)/(\gamma i + \delta)\} = (\alpha\delta - \beta\gamma)/(\gamma^2 + \delta^2) > 0$, the transformation maps the upper half-plane onto itself.

Now every mapping of the upper half-plane itself may be accomplished by a linear transformation which takes three real points x_1, x_2, x_3, $x_1 < x_2 < x_3$, into 0, 1, ∞, respectively. If we form such a linear function (see problem 12), we obtain a transformation of the form given above.

16. To find the mapping use the result of problem 12. This yields

$$\frac{w - 0}{w - \infty} \div \frac{1 - 0}{1 - \infty} = \frac{z - 1}{z + 1} \div \frac{i - 1}{i + 1},$$

or $w = -i(z - 1)/(z + 1)$. $z = 0$ is taken into $w = i$, $z = \infty$ into $w = -i$. It follows that the straight lines through the origin are taken into circles through i and $-i$, the radii into the arcs of these circles contained within the upper half-plane.

17. To find the mapping function use the result of problem 12. This yields $w = (z + i)/(z - i)$. Thus $z = 0$ is taken into $w = -1$, $z = \infty$ into $w = +1$. The straight lines through the origin are taken into circles passing through -1 and $+1$, the rays am $z =$ const., $|z| \geq 1$ into the arcs of these circles lying within the right half-plane. The circles $|z| = r > 1$ are taken into circles orthogonal to the circles containing $+1$ and -1 and lying within the right half-plane.

18. Let $w = l(z)$ be a linear mapping of the interior of the unit circle onto itself. Let a, $|a| < 1$ be the point which is taken into $w = 0$. It is easy to verify that the mapping

$$w = l_1(z) = \frac{z - a}{1 - \bar{a}z}$$

takes the unit circle into itself and $z = a$ into $w = 0$.

$l(z)$ differs from $l_1(z)$ at most by a linear transformation which leaves the unit circle and the origin fixed. Such a transformation leaves the point ∞ fixed (see problem 13); hence it is a rotation about the origin (see problem 11). It follows that $l(z) = e^{i\alpha}l_1(z)$, and that

$$w = e^{i\alpha}\,\frac{z - a}{1 - \bar{a}z}, \qquad \alpha \text{ real}, \qquad |\,a\,| < 1$$

is the most general mapping of the desired type. It may be written in the form $w = (az + b)/(\bar{b}z + \bar{a})$, $|\,a\,| > |\,b\,|$.

19. Every rotation of the sphere is a rotation about an axis. Let P and P' be the end-points of the axis. Their images in the plane are z_0 , $1/\bar{z}_0$) (see problem 9b). These two points are fixed points of the linear transformation. The result of problem 11 shows that the transformation is of the form

$$\frac{w - z_0}{w - 1/\bar{z}_0} = A \cdot \frac{z - z_0}{z - 1/\bar{z}_0}.$$

It is easy to see that the transformation must be elliptic (see, for instance, the answer to problem 21b). Setting $|\,A\,| = 1$ we obtain after a simple computation: $w = (az + b)/(-\bar{b}z + \bar{a})$, $a \neq 0$, b arbitrary.

20. The transformation is not uniquely determined. To find the transformation, draw the straight line joining the two centers. There exists on this line a uniquely determined pair of points z_0 , z_0' such that z_0' is the reflection of z_0 with respect to both circles k_1

and k_2. Any linear transformation taking z_0 into 0 and z_0' into ∞ (or z_0 into ∞ and z_0' into 0) takes k_1 and k_2 into two concentric circles about the origin.

21. a) $1/(w - \zeta_1) = 1/(z - \zeta_1) + C$, $C \neq 0$. If $\zeta_0 = \infty$, the transformation has the form $w = z + C$. Transformations with a single fixed point are called parabolic.

b) A hyperbolic transformation leaves each circle through ζ_1 and ζ_2 fixed, takes each orthogonal circle into another such circle. An elliptic transformation takes a circle through ζ_1 and ζ_2 into another such circle, leaves each orthogonal circle fixed. The behaviour of a loxodromic transformation can be described by representing it as an elliptic transformation followed by a hyperbolic one.

A parabolic transformation takes each circle through ζ_1 into a circle possessing at this point the same tangent. Circles tangent to one distinguished line remain fixed.

All this becomes obvious if we consider first the case $\zeta_1 = 0$, $\zeta_2 = \infty$ (or in the parabolic case $\zeta_1 = \infty$) and note that a linear transformation of both the z and the w planes permits us to put the fixed points wherever we please.

22. Assume first that the transformation possesses two fixed points ζ_1 and ζ_2. Set $a' = (a - c\zeta_1)/(a - c\zeta_2)$. If $|a'| \neq 0$ (hyperbolic or laxodromic transformation), then $z_\nu \to \zeta_1$ if $|a'| < 1$, $z_\nu \to \zeta_2$ if $|a'| > 1$. If $|a'| = 1$ (elliptic transformation), then all points z_ν lie on a circle. The sequence (z_ν) does not converge. It contains infinitely many distinct points if and only if $(\text{am } z_0)/2\pi$ is irrational.

If the transformation possesses a single fixed point

ζ_1 (parabolic case), then $z_\nu \to \zeta_1$. (We have neglected the trivial cases $z_0 = \zeta_1$, $z_0 = \zeta_2$.)

§13. Simple Non-Linear Mapping Problems

1. We have $|e^z| = |e^{x+iy}| = e^x$, am $e^z = y$. The image of the segment $x = x_0 - \pi, < y \leq +\pi$ is the circle $|w| = e^x$. When y increases from $-\pi$ (exclusive) to $+\pi$ (inclusive), the circle is traversed once in the positive direction. When x_0 increases from $-\infty$ to $+\infty$, the radius of the circle increases from $-\infty$ (exclusive) to $+\infty$ (exclusive). The image of the straight line $-\infty < x < +\infty$, $y = y_0$, is the ray am $w = y_0$. When x increases from $-\infty$ to $+\infty$, this ray is traversed once from 0 (exclusive) to ∞ (exclusive). When y_0 increases from $-\pi$ to $+\pi$, the ray is rotated in the clockwise direction from the negative real axis (exclusive) to the negative real axis (inclusive).

2. We have

$$w = u + iv = \sin z = \sin (x + iy)$$

$$= \sin x \, \frac{e^y + e^{-y}}{2} + i \cos x \, \frac{e^y - e^{-y}}{2},$$

i.e.

$$u = \sin x \, \frac{e^y + e^{-y}}{2}, \qquad v = \cos x \, \frac{e^y - e^{-y}}{2}.$$

Let y_0 be a fixed positive number. If $y = y_0$ and x increases from $-\pi$ to $+\pi$, then w describes an ellipse with foci at ± 1. The ellipse is described once in the clockwise direction starting at its lowest point. For $y = -y_0$ we obtain the same ellipse described in the opposite direction starting at its highest point. Through each point of the plane (except those along the real segment $-1 \cdots +1$) passes exactly one such ellipse. If $y_0 = 0$, we describe the segment $-1 \cdots +1$ twice (from 0 to -1, to 0, to $+1$ to 0). In order to obtain the complete image of the period strip, we take two copies of the w-plane, cut one along the positive imaginary axis, the other along the negative imaginary axis and both along the real segment $-1 \cdots +1$. If we join the banks of the horizontal slits crosswise, we obtain the two-sheeted Riemann surface which is the one-to-one image of the period strips.

The lines $\Re(z) = $ const. are taken into hyperbolas orthogonal to the ellipses.

3. The best way to visualize this mapping is to proceed step-by-step. Set $2z = z_1$, $iz_1 = z_2$, $e^{z_2} = z_3$, $(z_3 - 1)/(z_3 + 1) = z_{4_1}$, $-iz_4 = \tan z = w$. The similarity transformation from the z-plane to the z_2-plane takes our strip, segments and straight lines into the strip, segments and straight lines of problem 1. In the z_3-plane we obtain the rays through the origin and the circles about the origin. The linear transformation from the z_3-plane to the z_4-plane yields the circles through $+1$ and -1 and the orthogonal circles. The transformation from the z_4-plane to the w-plane is a mere rotation by $-90°$ which takes $+1$ and -1 into $-i$ and $+i$.

4. Since sh $z = -i$ siniz, we obtain all desired information from the answer to problem 2.

5. Set $z = r(\cos \phi + i \sin \phi)$. Then

$$w = u + iv = \left(r + \frac{1}{r}\right) \cos\phi + \left(r - \frac{1}{r}\right) \sin\phi.$$

If $|z| = r > 1$ is fixed and ϕ increases from 0 to 2π, w traverses (in the counter-clockwise direction) an ellipse with foci at ± 2 and with semi-axes $r \pm 1/r$. If r takes on all values > 1, these ellipses fill out the region exterior to the segment $-2 \cdots 2$ except the point $w = \infty$. This is the image of the region $1 < |z| < \infty$.

6. The image of the circle $|z| = 1/r$, $r > 1$, is the same ellipse which was obtained in the preceding problem (traversed in the clockwise direction). The image of the region $0 < |z| < 1$ is identical with that of the region $1 < |z| < +\infty$. To obtain the image of the region $0 < |z| + \infty$, take two copies of the w-plane cut along the segments $-2 \cdots 2$ and join the banks of the cuts crosswise. The resulting Riemann surface is doubly-connected; the boundary consists of the points $w = \infty$ on the two sheets.

7. a) The closed upper half of the region bounded by the ellipse with foci at ± 2 and semi-axis $5/2$, $3/2$.

b) The closed lower half of the same region.

c) The image consists of the domains described under a) and b) joined along the segment $-2 \cdots 2$. Its boundary consists of the ellipse and the upper and lower banks of the slits $-5/2 \cdots 2$, $2 \cdots 5/2$.

8. If z traverses a ray (am $z = $ const.) from 0 to ∞, then w traverses one branch of the hyperbola

$$\left(\frac{u}{2\cos\phi}\right)^2 - \left(\frac{v}{2\sin\phi}\right)^2 = 1, \quad \phi = am\ z.$$

from ∞ to ∞. (Note the simple degenerate cases $\phi = 0$, $\pi/2$, π, $3\pi/2$.) For $0 < \text{am } z \leq \pi/3$, $|z| \geq 1$, we obtain the branch situated in the first quadrant and traversed starting from the point on the real axis. These considerations imply the following answers. In all three cases we deal with the hyperbola with foci at ± 2 and semi-axis 1 and 3.

a) The intersection of the closed region bounded by the right branch of the hyperbola and the (closed) first quadrant.

b) The closed region bounded by the right branch of the hyperbola cut along the positive real axis from 2 to $+\infty$. The upper bank of the cut corresponds to the part of the real axis from 1 to ∞, the lower bank to the segment $0 \cdots 1$.

c) The same region as in b) cut along the segment $+2 \cdots +1$.

9. Using the answer to problem 6 we see that the function

$$w = \frac{z - (z^2 - 4)^{1/2}}{2}$$

(obtained by solving the equation $z = w + 1/w$) maps the exterior of the ellipse onto the region $|w| < 1/7$. The desired mapping function is

$$w = \frac{7}{2} [z - (z^2 - 4)^{1/2}].$$

The point $z = \infty$ is taken into $w = 0$. Is the mapping conformal at $z = \infty$?

10. The mapping $z_1 = z^3$ takes the sector into the upper half of the unit disc (in the z_1-plane). The

mapping $z_2 = (1 + z_1)/(1 - z_1)$ yields the first quadrant (in the z_2-plane), the mapping $z_3 = z_2^2$, the upper half-plane (in the z_3-plane). The upper half-plane is taken into the unit disc by the transformation $w = (z_3 - i)/(z_3 + i)$. The desired mapping function is

$$w = \frac{(1 + z^3)^2 - i(1 - z^3)^2}{(1 + z^3)^2 + i(1 - z^3)^2}.$$

At which boundary points is the mapping conformal? Investigate the behaviour of the radii and the circular arcs $|z| = $ const. under the intermediate transformation used to obtain the final mapping.

11. a) The "vertices" of our region are $\rho = 1/2 - i3^{1/2}/2$, $\rho' = 1/2 + i3^{1/2}/2 = 1/\rho$. The mapping $z_1 = (z - \rho)/(1 - \rho z)$ yields the sector $0 \leq$ am $z \leq 2\pi/3$. The mapping $z_2 = (z_1^3)^{1/2}$ takes this sector into the upper half-plane. (The sign of the square-root is determined by the condition am $z_2 = (3/2)$ am z_1). The upper half-plane is taken into the unit disc by the transformation $w = (z_2 - i)/(z_2 + i)$. The desired mapping function is

$$w = \frac{\left(\dfrac{z - \rho}{1 - \rho z}\right)^{3/2} - i}{\left(\dfrac{z - \rho}{1 - \rho z}\right)^{3/2} + i}.$$

b) $w = \{(1 + z)^2 - i(1 - z)^2\}/\{(1 + z)^2 + i(1 - z)^2\}$. This mapping function may be constructed in the manner described in the answer to problem 10.

12. The mapping $z_1 = 1/(z - 1)$ yields the strip

$-1 \leq \Re(z_1) \leq -1/2$, the mapping $z_2 = 2\pi i(z_1 + 1)$ the strip $0 \leq \Im(z_2) \leq +\pi$, the mapping $z_3 = e^{z_2}$ the upper half-plane $\Im(z_3) \geq 0$ (see problem 1), the mapping $w = (z_3 - i)/(z_3 + i)$ the unit disc. Investigate the transformations experienced by the circles tangent to the given ones at $z = 1$ and by their orthogonal circles.

13. The mapping $z_1 = (z + 1/2)\pi i$ yields the half-strip $\Re(z_1) \leq 0$, $0 \leq \Im(z) \leq \pi$, the mapping $z_2 = e^{z_1}$ the upper half of the unit disc in the z_2-plane, the mapping $z_3 = (1 + z_2)/(1 - z_2)$ (see problem 10) the first quadrant of the z_3-plane, the mapping $z_4 = z_3^2$ the upper half-plane $\Im(z_4) \geq 0$, the mapping $w = (z_4 - i)/(z_4 + i)$ the unit disc. The desired mapping function is

$$\frac{(1 + e^{(z+1/2)i\pi})^2 - i(1 - e^{(z+1/2)i\pi})^2}{(1 + e^{(z+1/2)i\pi})^2 + i(1 - e^{(z+1/2)i\pi})^2}.$$

Investigate the transformation experienced by the rays and segments which are orthogonal to the boundary lines of the half-strip in the z-plane.

14. The equation of the parabola is $y^2 = 4\alpha^2(x + \alpha^2)$. At first we cut our region along the axis of the parabola from 0 to $+\infty$ and denote upper and lower banks of the cut by s_1 and s_2, respectively. Later we will have to join s_1 and s_2 point-by-point. We denote the parabola by p, the segment $-\alpha^2 \cdots 0$ by t. The mapping $z_1 = z^{1/2}$ takes the cut region into the strip $0 \leq \Im(z_1) \leq \alpha$, provided we use an appropriate branch of the square root. The mapping $z_2 = \pi z_1/\alpha$ yields the strip $0 \leq \Im(z_2) \leq \pi$. s_1 is taken into the positive real axis, s_2 into the negative real axis, p

into the remaining part of the boundary, t into a segment along the imaginary axis. Setting $z_3 = e^{z_2}$, we obtain the upper half-plane $\Im(z_3) \geq 0$ (see problem 1). s_1 is taken into the real ray $+1 \cdots +\infty$, s_2 into the segment $1 \cdots 0$, p into the negative real axis, t into the upper half of the unit circle. We can not apply directly the transformation $(z_3 - i)/(z_3 + i)$ in order to obtain the unit disc, for this transformation would separate s_1 and s_2. Instead we map the upper half-plane of the z_3-plane onto the upper half-plane $\Im(z_4) \geq 0$ by the transformation $z_4 = (z_3 - 1)/(z_3 + 1)$. s_1 and s_2 are taken into the segments $0 \cdots +1$, $0 \cdots -1$, p into the remaining part of the real axis, t into the imaginary axis. The mapping $z_5 = z_4^2$ joins s_1 with s_2 and takes p into the upper and lower banks of the cut extending along the positive real axis from $+1$ to ∞. The mapping $z_6 = z_5 - 1$ gives us the entire plane cut along the positive real axis as an image of the original domain. The two banks of the cut correspond to the two branches of the parabola. To separate these two banks we use the transformation $z_7 = z_6^{1/2}$. If an appropriate branch of the root is used, we obtain the upper half-plane $\Im(z_7) \geq 0$; its boundary is the image of p. (Where are the images of t, s_1, s_2?) Now we may pass to the unit disc by setting $w = (z_7 - i)/(z_7 + i)$. Combining these transformations, we obtain the desired mapping function in the form

$$w = \tan^2\left(\frac{\pi}{4\alpha i} z^{1/2}\right).$$

15. a) If $\Re(z) = 1/2$, then z is conjugate to $1 - z$, so that $|\log z| = |\log (1 - z)|$. This proves the

"middle part" of our assertion. Next, we consider the function

$$f(z) = \frac{\log z}{\log (1 - z)}$$

in the quadrant $\Re(z) > 1/2$, $\Im(z) > 0$. $f(z)$ is regular in the interior and on the boundary of this quadrant, except at $z = 1$ and $z = \infty$. On the vertical boundary $|f(z)| = 1$. On the horizontal boundary $z = x > 1/2$. Hence $|f(z)| < 1$ on this boundary. For $1/2 < x < 1$ this is obvious, for $x > 1$ this follows from the inequality $\log^2 x - \log^2(x - 1) < \pi^2$ which implies that

$$\left| \frac{\log x}{\log (1 - x)} \right| = \left| \frac{\log x}{\log (x - 1) + i\pi} \right| < 1.$$

If z lies within the closed quadrant and $z \to +1$ or $z \to +\infty$, then $|f(z)| \to 0$ and $|f(z)| \to 1$, respectively. Now we draw a circle of radius $\rho < 1/2$ about $+1$ and a circle of radius $R > 2$ about $+ 1/2$. Let $\mathfrak{G}$ denote the part of our quadrant enclosed between the two circles. The boundary of $\mathfrak{G}$ consists of an arc of the smaller circle, an arc of the larger circle, and two straight lines. $f(z)$ is regular in $\mathfrak{G}$ and assumes its maximum on the boundary (KI, p. 85). On the vertical boundary $|f(z)| = 1$, on the horizontal $|f(z)| < 1$, on the arc of the smaller circle the maximum of $|f(z)|$ will be less than 1 if we choose ρ sufficiently small. On the arc of the larger circle this maximum will be arbitrarily close to 1 if we choose R sufficiently large. It follows that the maximum of $|f(z)|$ in G is arbitrarily close to 1. Thus $|f(z)| \leq 1$ in $\mathfrak{G}$, and since f

is not constant $|f(z)| < 1$. This inequality holds at all interior points of our quadrant. In a similar way, we can show that $|f(z)| < 1$ for $\Re(z) > 1/2$, $\Im(z) < 0$. Setting $|f(0)| = 1$, we see that $|f(z)| < 1$ whenever $\Re(z) > 1/2$. Replacing z by $1 - z$ and setting $|f(0)| = +\infty$, we see that $|f(z)| > 1$ whenever $\Re(z) < 1/2$. This completes the proof of a).

b) follows from a) by replacing z by $1/z$; c) follows from a) by replacing z by $1/(1 - z)$.

16. Consider the domains

$$\mathfrak{E}: |z| < 1, \qquad \mathfrak{K}: |z - 1| < 1$$

and denote by $\mathfrak{A}$ the intersection of $\mathfrak{K}$ with the half-plane $\Re(z) > 1/2$, by $\mathfrak{B}$ the intersection of $\mathfrak{E}$ with the half-plane $\Re(z) < 1/2$ and by $\mathfrak{C}$ the region defined by the inequalities $|z| > 1$, $|z - 1| > 1$. Of our three numbers, the first is the smallest in $\mathfrak{A}$, the second in $\mathfrak{B}$, the third in $\mathfrak{C}$. The proof follows from the result of problem 15.

Where are two of the numbers equal? Where are all three equal?

The boundaries of $\mathfrak{E}$ and $\mathfrak{K}$ and the straight line $\Re(z) = 1/2$ divide the plane into 6 regions. In each of these regions determine the inequalities satisfied by our three numbers.

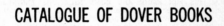
CATALOGUE OF DOVER BOOKS

MATHEMATICS—INTERMEDIATE TO ADVANCED

General

INTRODUCTION TO APPLIED MATHEMATICS, Francis D. Murnaghan. A practical and thoroughly sound introduction to a number of advanced branches of higher mathematics. Among the selected topics covered in detail are: vector and matrix analysis, partial and differential equations, integral equations, calculus of variations, Laplace transform theory, the vector triple product, linear vector functions, quadratic and bilinear forms, Fourier series, spherical harmonics, Bessel functions, the Heaviside expansion formula, and many others. Extremely useful book for graduate students in physics, engineering, chemistry, and mathematics. Index. 111 study exercises with answers. 41 illustrations. ix + .389pp. 5⅜ x 8½.
S1042 Paperbound **$2.00**

OPERATIONAL METHODS IN APPLIED MATHEMATICS, H. S. Carslaw and J. C. Jaeger. Explanation of the application of the Laplace Transformation to differential equations, a simple and effective substitute for more difficult and obscure operational methods. Of great practical value to engineers and to all workers in applied mathematics. Chapters on: Ordinary Linear Differential Equations with Constant Coefficients;; Electric Circuit Theory; Dynamical Applications; The Inversion Theorem for the Laplace Transformation; Conduction of Heat; Vibrations of Continuous Mechanical Systems; Hydrodynamics; Impulsive Functions; Chains of Differential Equations; and other related matters. 3 appendices. 153 problems, many with answers. 22 figures. xvi + 359pp. 5⅜ x 8½.
S1011 Paperbound **$2.25**

APPLIED MATHEMATICS FOR RADIO AND COMMUNICATIONS ENGINEERS, C. E. Smith. No extraneous material here!—only the theories, equations, and operations essential and immediately useful for radio work. Can be used as refresher, as handbook of applications and tables, or as full home-study course. Ranges from simplest arithmetic through calculus, series, and wave forms, hyperbolic trigonometry, simultaneous equations in mesh circuits, etc. Supplies applications right along with each math topic discussed. 22 useful tables of functions, formulas, logs, etc. Index. 166 exercises, 140 examples, all with answers. 95 diagrams. Bibliography. x + 336pp. 5⅜ x 8.
S141 Paperbound **$1.75**

Algebra, group theory

ALGEBRAS AND THEIR ARITHMETICS, L. E. Dickson. Provides the foundation and background necessary to any advanced undergraduate or graduate student studying abstract algebra. Begins with elementary introduction to linear transformations, matrices, field of complex numbers; proceeds to order, basal units, modulus, quaternions, etc.; develops calculus of linears sets, describes various examples of algebras including invariant, difference, nilpotent, semi-simple. "Makes the reader marvel at his genius for clear and profound analysis," Amer. Mathematical Monthly. Index. xii + 241pp. 5⅜ x 8.
S616 Paperbound **$1.50**

THE THEORY OF EQUATIONS WITH AN INTRODUCTION TO THE THEORY OF BINARY ALGEBRAIC FORMS, W. S. Burnside and A. W. Panton. Extremely thorough and concrete discussion of the theory of equations, with extensive detailed treatment of many topics curtailed in later texts. Covers theory of algebraic equations, properties of polynomials, symmetric functions, derived functions, Horner's process, complex numbers and the complex variable, determinants and methods of elimination, invariant theory (nearly 100 pages), transformations, introduction to Galois theory, Abelian equations, and much more. Invaluable supplementary work for modern students and teachers. 759 examples and exercises. Index in each volume. Two volume set. Total of xxiv + 604pp. 5⅜ x 8.
S714 Vol I Paperbound **$1.85**
S715 Vol II Paperbound **$1.85**
The set **$3.70**

COMPUTATIONAL METHODS OF LINEAR ALGEBRA, V. N. Faddeeva, translated by **C. D. Benster.** First English translation of a unique and valuable work, the only work in English presenting a systematic exposition of the most important methods of linear algebra—classical and contemporary. Shows in detail how to derive numerical solutions of problems in mathematical physics which are frequently connected with those of linear algebra. Theory as well as individual practice. Part I surveys the mathematical background that is indispensable to what follows. Parts II and III, the conclusion, set forth the most important methods of solution, for both exact and iterative groups. One of the most outstanding and valuable features of this work is the 23 tables, double and triple checked for accuracy. These tables will not be found elsewhere. Author's preface. Translator's note. New bibliography and index. x + 252pp. 5⅜ x 8.
S424 Paperbound **$1.95**

ALGEBRAIC EQUATIONS, E. Dehn. Careful and complete presentation of Galois' theory of algebraic equations; theories of Lagrange and Galois developed in logical rather than historical form, with a more thorough exposition than in most modern books. Many concrete applications and fully-worked-out examples. Discusses basic theory (very clear exposition of the symmetric group); isomorphic, transitive, and Abelian groups; applications of Lagrange's and Galois' theories; and much more. Newly revised by the author. Index. List of Theorems. xi + 208pp. 5⅜ x 8.
S697 Paperbound **$1.45**

THEORY AND APPLICATIONS OF FINITE GROUPS, G. A. Miller, H. F. Blichfeldt, L. E. Dickson. Unusually accurate and authoritative work, each section prepared by a leading specialist: Miller on substitution and abstract groups, Blichfeldt on finite groups of linear homogeneous transformations, Dickson on applications of finite groups. Unlike more modern works, this gives the concrete basis from which abstract group theory arose. Includes Abelian groups, prime-power groups, isomorphisms, matrix forms of linear transformations, Sylow groups, Galois' theory of algebraic equations, duplication of a cube, trisection of an angle, etc. 2 Indexes. 267 problems. xvii + 390pp. 5⅜ x 8. S216 Paperbound **$2.00**

THE THEORY OF DETERMINANTS, MATRICES, AND INVARIANTS, H. W. Turnbull. Important study includes all salient features and major theories. 7 chapters on determinants and matrices cover fundamental properties, Laplace identities, multiplication, linear equations, rank and differentiation, etc. Sections on invariants gives general properties, symbolic and direct methods of reduction, binary and polar forms, general linear transformation, first fundamental theorem, multilinear forms. Following chapters study development and proof of Hilbert's Basis Theorem, Gordan-Hilbert Finiteness Theorem, Clebsch's Theorem, and include discussions of apolarity, canonical forms, geometrical interpretations of algebraic forms, complete system of the general quadric, etc. New preface and appendix. Bibliography. xviii + 374pp. 5⅜ x 8. S699 Paperbound **$2.25**

AN INTRODUCTION TO THE THEORY OF CANONICAL MATRICES, H. W. Turnbull and A. C. Aitken. All principal aspects of the theory of canonical matrices, from definitions and fundamental properties of matrices to the practical applications of their reduction to canonical form. Beginning with matrix multiplications, reciprocals, and partitioned matrices, the authors go on to elementary transformations and bilinear and quadratic forms. Also covers such topics as a rational canonical form for the collineatory group, congruent and conjunctive transformation for quadratic and hermitian forms, unitary and orthogonal transformations, canonical reduction of pencils of matrices, etc. Index. Appendix. Historical notes at chapter ends. Bibliographies. 275 problems. xiv + 200pp. 5⅜ x 8. S177 Paperbound **$1.55**

A TREATISE ON THE THEORY OF DETERMINANTS, T. Muir. Unequalled as an exhaustive compilation of nearly all the known facts about determinants up to the early 1930's. Covers notation and general properties, row and column transformation, symmetry, compound determinants, adjugates, rectangular arrays and matrices, linear dependence, gradients, Jacobians, Hessians, Wronskians, and much more. Invaluable for libraries of industrial and research organizations as well as for student, teacher, and mathematician; very useful in the field of computing machines. Revised and enlarged by W. H. Metzler. Index. 485 problems and scores of numerical examples. iv + 766pp. 5⅜ x 8. S670 Paperbound **$3.00**

THEORY OF DETERMINANTS IN THE HISTORICAL ORDER OF DEVELOPMENT, Sir Thomas Muir. Unabridged reprinting of this complete study of 1,859 papers on determinant theory written between 1693 and 1900. Most important and original sections reproduced, valuable commentary on each. No other work is necessary for determinant research: all types are covered—each subdivision of the theory treated separately; all papers dealing with each type are covered; you are told exactly what each paper is about and how important its contribution is. Each result, theory, extension, or modification is assigned its own identifying numeral so that the full history may be more easily followed. Includes papers on determinants in general, determinants and linear equations, symmetric determinants, alternants, recurrents, determinants having invariant factors, and all other major types. "A model of what such histories ought to be," NATURE. "Mathematicians must ever be grateful to Sir Thomas for his monumental work," AMERICAN MATH MONTHLY. Four volumes bound as two. Indices. Bibliographies. Total of lxxxiv + 1977pp. 5⅜ x 8. S672-3 The set, Clothbound **$12.50**

Calculus and function theory, Fourier theory, infinite series, calculus of variations, real and complex functions

FIVE VOLUME "THEORY OF FUNCTIONS' SET BY KONRAD KNOPP

This five-volume set, prepared by Konrad Knopp, provides a complete and readily followed account of theory of functions. Proofs are given concisely, yet without sacrifice of completeness or rigor. These volumes are used as texts by such universities as M.I.T., University of Chicago, N. Y. City College, and many others. "Excellent introduction . . . remarkably readable, concise, clear, rigorous," JOURNAL OF THE AMERICAN STATISTICAL ASSOCIATION.

ELEMENTS OF THE THEORY OF FUNCTIONS, Konrad Knopp. This book provides the student with background for further volumes in this set, or texts on a similar level. Partial contents: foundations, system of complex numbers and the Gaussian plane of numbers, Riemann sphere of numbers, mapping by linear functions, normal forms, the logarithm, the cyclometric functions and binomial series. "Not only for the young student, but also for the student who knows all about what is in it," MATHEMATICAL JOURNAL. Bibliography. Index. 140pp. 5⅜ x 8. S154 Paperbound **$1.35**

THEORY OF FUNCTIONS, PART I, Konrad Knopp. With volume II, this book provides coverage of basic concepts and theorems. Partial contents: numbers and points, functions of a complex variable, integral of a continuous function, Cauchy's integral theorem, Cauchy's integral formulae, series with variable terms, expansion of analytic functions in power series, analytic continuation and complete definition of analytic functions, entire transcendental functions, Laurent expansion, types of singularities. Bibliography. Index. vii + 146pp. 5⅜ x 8. S156 Paperbound **$1.35**

THEORY OF FUNCTIONS, PART II, Konrad Knopp. Application and further development of general theory, special topics. Single valued functions, entire, Weierstrass, Meromorphic functions. Riemann surfaces. Algebraic functions. Analytical configuration, Riemann surface. Bibliography. Index. x + 150pp. 5⅜ x 8. S157 Paperbound **$1.35**

PROBLEM BOOK IN THE THEORY OF FUNCTIONS, VOLUME 1, Konrad Knopp. Problems in elementary theory, for use with Knopp's THEORY OF FUNCTIONS, or any other text, arranged according to increasing difficulty. Fundamental concepts, sequences of numbers and infinite series, complex variable, integral theorems, development in series, conformal mapping. 182 problems. Answers. viii + 126pp. 5⅜ x 8. S158 Paperbound **$1.35**

PROBLEM BOOK IN THE THEORY OF FUNCTIONS, VOLUME 2, Konrad Knopp. Advanced theory of functions, to be used either with Knopp's THEORY OF FUNCTIONS, or any other comparable text. Singularities, entire & meromorphic functions, periodic, analytic, continuation, multiple-valued functions, Riemann surfaces, conformal mapping. Includes a section of additional elementary problems. "The difficult task of selecting from the immense material of the modern theory of functions the problems just within the reach of the beginner is here masterfully accomplished," AM. MATH. SOC. Answers. 138pp. 5⅜ x 8. S159 Paperbound **$1.35**

A COURSE IN MATHEMATICAL ANALYSIS, Edouard Goursat. Trans. by E. R. Hedrick, O. Dunkel. Classic study of fundamental material thoroughly treated. Exceptionally lucid exposition of wide range of subject matter for student with 1 year of calculus. Vol. 1: Derivatives and Differentials, Definite Integrals, Expansion in Series, Applications to Geometry. Problems. Index. 52 illus. 556pp. Vol. 2, Part I: Functions of a Complex Variable, Conformal Representations, Doubly Periodic Functions, Natural Boundaries, etc. Problems. Index. 38 illus. 269pp. Vol. 2, Part 2: Differential Equations, Cauchy-Lipschitz Method, Non-linear Differential Equations, Simultaneous Equations, etc. Problems. Index. 308pp. 5⅜ x 8.
Vol. 1 S554 Paperbound **$2.50**
Vol. 2 part 1 S555 Paperbound **$1.85**
Vol. 2 part 2 S556 Paperbound **$1.85**
3 vol. set **$6.20**

MODERN THEORIES OF INTEGRATION, H. Kestelman. Connected and concrete coverage, with fully-worked-out proofs for every step. Ranges from elementary definitions through theory of aggregates, sets of points, Riemann and Lebesgue integration, and much more. This new revised and enlarged edition contains a new chapter on Riemann-Stieltjes integration, as well as a supplementary section of 186 exercises. Ideal for the mathematician, student, teacher, or self-studier. Index of Definitions and Symbols. General Index. Bibliography. x + 310pp. 5⅝ x 8⅜. S572 Paperbound **$2.25**

THEORY OF MAXIMA AND MINIMA, H. Hancock. Fullest treatment ever written; only work in English with extended discussion of maxima and minima for functions of 1, 2, or n variables, problems with subsidiary constraints, and relevant quadratic forms. Detailed proof of each important theorem. Covers the Scheeffer and von Dantscher theories, homogeneous quadratic forms, reversion of series, fallacious establishment of maxima and minima, etc. Unsurpassed treatise for advanced students of calculus, mathematicians, economists, statisticians. Index. 24 diagrams. 39 problems, many examples. 193pp. 5⅜ x 8. S665 Paperbound **$1.50**

AN ELEMENTARY TREATISE ON ELLIPTIC FUNCTIONS, A. Cayley. Still the fullest and clearest text on the theories of Jacobi and Legendre for the advanced student (and an excellent supplement for the beginner). A masterpiece of exposition by the great 19th century British mathematician (creator of the theory of matrices and abstract geometry), it covers the addition-theory, Landen's theorem, the 3 kinds of elliptic integrals, transformations, the q-functions, reduction of a differential expression, and much more. Index. xii + 386pp. 5⅜ x 8. S728 Paperbound **$2.00**

THE APPLICATIONS OF ELLIPTIC FUNCTIONS, A. G. Greenhill. Modern books forego detail for sake of brevity—this book offers complete exposition necessary for proper understanding, use of elliptic integrals. Formulas developed from definite physical, geometric problems; examples representative enough to offer basic information in widely useable form. Elliptic integrals, addition theorem, algebraical form of addition theorem, elliptic integrals of 2nd, 3rd kind, double periodicity, resolution into factors, series, transformation, etc. Introduction. Index. 25 illus. xi + 357pp. 5⅜ x 8. S603 Paperbound **$1.75**

THE THEORY OF FUNCTIONS OF REAL VARIABLES, James Pierpont. A 2-volume authoritative exposition, by one of the foremost mathematicians of his time. Each theorem stated with all conditions, then followed by proof. No need to go through complicated reasoning to discover conditions added without specific mention. Includes a particularly complete, rigorous presentation of theory of measure; and Pierpont's own work on a theory of Lebesgue integrals, and treatment of area of a curved surface. Partial contents, Vol. 1: rational numbers, exponentials, logarithms, point aggregates, maxima, minima, proper integrals, improper integrals, multiple proper integrals, continuity, discontinuity, indeterminate forms. Vol. 2: point sets, proper integrals, series, power series, aggregates, ordinal numbers, discontinuous functions, sub-, infra-uniform convergence, much more. Index. 95 illustrations. 1229pp. 5⅜ x 8. S558-9, 2 volume set, paperbound **$5.20**

FUNCTIONS OF A COMPLEX VARIABLE, James Pierpont. Long one of best in the field. A thorough treatment of fundamental elements, concepts, theorems. A complete study, rigorous, detailed, with carefully selected problems worked out to illustrate each topic. Partial contents: arithmetical operations, real term series, positive term series, exponential functions, integration, analytic functions, asymptotic expansions, functions of Weierstrass, Legendre, etc. Index. List of symbols. 122 illus. 597pp. 5⅜ x 8. S560 Paperbound **$2.45**

MODERN OPERATIONAL CALCULUS: WITH APPLICATIONS IN TECHNICAL MATHEMATICS, N. W. McLachlan. An introduction to modern operational calculus based upon the Laplace transform, applying it to the solution of ordinary and partial differential equations. For physicists, engineers, and applied mathematicians. Partial contents: Laplace transform, theorems or rules of the operational calculus, solution of ordinary and partial linear differential equations with constant coefficients, evaluation of integrals and establishment of mathematical relationships, derivation of Laplace transforms of various functions, etc. Six appendices deal with Heaviside's unit function, etc. Revised edition. Index. Bibliography. xiv + 218pp. 5⅜ x 8½. S192 Paperbound **$1.75**

ADVANCED CALCULUS, E. B. Wilson. An unabridged reprinting of the work which continues to be recognized as one of the most comprehensive and useful texts in the field. It contains an immense amount of well-presented, fundamental material, including chapters on vector functions, ordinary differential equations, special functions, calculus of variations, etc., which are excellent introductions to these areas. For students with only one year of calculus, more than 1300 exercises cover both pure math and applications to engineering and physical problems. For engineers, physicists, etc., this work, with its 54 page introductory review, is the ideal reference and refresher. Index. ix + 566pp. 5⅜ x 8. S504 Paperbound **$2.45**

ASYMPTOTIC EXPANSIONS, A. Erdélyi. The only modern work available in English, this is an unabridged reproduction of a monograph prepared for the Office of Naval Research. It discusses various procedures for asymptotic evaluation of integrals containing a large parameter and solutions of ordinary linear differential equations. Bibliography of 71 items. vi + 108pp. 5⅜ x 8. S318 Paperbound **$1.35**

INTRODUCTION TO ELLIPTIC FUNCTIONS: with applications, F. Bowman. Concise, practical introduction to elliptic integrals and functions. Beginning with the familiar trigonometric functions, it requires nothing more from the reader than a knowledge of basic principles of differentiation and integration. Discussion confined to the Jacobian functions. Enlarged bibliography. Index. 173 problems and examples. 56 figures, 4 tables. 115pp. 5⅜ x 8. S922 Paperbound **$1.25**

ON RIEMANN'S THEORY OF ALGEBRAIC FUNCTIONS AND THEIR INTEGRALS: A SUPPLEMENT TO THE USUAL TREATISES, Felix Klein. Klein demonstrates how the mathematical ideas in Riemann's work on Abelian integrals can be arrived at by thinking in terms of the flow of electric current on surfaces. Intuitive explanations, not detailed proofs given in an extremely clear exposition, concentrating on the kinds of functions which can be defined on Riemann surfaces. Also useful as an introduction to the origins of topological problems. Complete and unabridged. Approved translation by Frances Hardcastle. New introduction. 43 figures. Glossary. xii + 76pp. 5⅜ x 8½. S1072 Paperbound **$1.25**

COLLECTED WORKS OF BERNHARD RIEMANN. This important source book is the first to contain the complete text of both 1892 Werke and the 1902 supplement, unabridged. It contains 31 monographs, 3 complete lecture courses, 15 miscellaneous papers, which have been of enormous importance in relativity, topology, theory of complex variables, and other areas of mathematics. Edited by R. Dedekind, H. Weber, M. Noether, W. Wirtinger. German text. English introduction by Hans Lewy. 690pp. 5⅜ x 8. S226 Paperbound **$3.75**

THE TAYLOR SERIES, AN INTRODUCTION TO THE THEORY OF FUNCTIONS OF A COMPLEX VARIABLE, P. Dienes. This book investigates the entire realm of analytic functions. Only ordinary calculus is needed, except in the last two chapters. Starting with an introduction to real variables and complex algebra, the properties of infinite series, elementary functions, complex differentiation and integration are carefully derived. Also biuniform mapping, a thorough two part discussion of representation and singularities of analytic functions, overconvergence and gap theorems, divergent series, Taylor series on its circle of convergence, divergence and singularities, etc. Unabridged, corrected reissue of first edition. Preface and index. 186 examples, many fully worked out. 67 figures. xii + 555pp. 5⅜ x 8. S391 Paperbound **$2.75**

INTRODUCTION TO BESSEL FUNCTIONS, Frank Bowman. A rigorous self-contained exposition providing all necessary material during the development, which requires only some knowledge of calculus and acquaintance with differential equations. A balanced presentation including applications and practical use. Discusses Bessel Functions of Zero Order, of Any Real Order; Modified Bessel Functions of Zero Order; Definite Integrals; Asymptotic Expansions; Bessel's Solution to Kepler's Problem; Circular Membranes; much more. "Clear and straightforward . . . useful not only to students of physics and engineering, but to mathematical students in general," Nature. 226 problems. Short tables of Bessel functions. 27 figures. Index. x + 135pp. 5⅜ x 8. S462 Paperbound **$1.35**

ELEMENTS OF THE THEORY OF REAL FUNCTIONS, J. E. Littlewood. Based on lectures given at Trinity College, Cambridge, this book has proved to be extremely successful in introducing graduate students to the modern theory of functions. It offers a full and concise coverage of classes and cardinal numbers, well-ordered series, other types of series, and elements of the theory of sets of points. 3rd revised edition. vii + 71pp. 5⅜ x 8.
S171 Clothbound **$2.85**
S172 Paperbound **$1.25**

TRANSCENDENTAL AND ALGEBRAIC NUMBERS, A. O. Gelfond. First English translation of work by leading Soviet mathematician. Thue-Siegel theorem, its p-adic analogue, on approximation of algebraic numbers by numbers in fixed algebraic field; Hermite-Lindemann theorem on transcendency of Bessel functions, solutions of other differential equations; Gelfond-Schneider theorem on transcendence of alpha to power beta; Schneider's work on elliptic functions, with method developed by Gelfond. Translated by L. F. Boron. Index. Bibliography. 200pp. 5⅜ x 8.
S615 Paperbound **$1.75**

ELLIPTIC INTEGRALS, H. Hancock. Invaluable in work involving differential equations containing cubics or quartics under the root sign, where elementary calculus methods are inadequate. Practical solutions to problems that occur in mathematics, engineering, physics: differential equations requiring integration of Lamé's, Briot's, or Bouquet's equations; determination of arc of ellipse, hyperbola, lemniscate; solutions of problems in elastica; motion of a projectile under resistance varying as the cube of the velocity; pendulums; many others. Exposition is in accordance with Legendre-Jacobi theory and includes rigorous discussion of Legendre transformations. 20 figures. 5 place table. Index. 104pp. 5⅛ x 8.
S484 Paperbound **$1.25**

LECTURES ON THE THEORY OF ELLIPTIC FUNCTIONS, H. Hancock. Reissue of the only book in English with so extensive a coverage, especially of Abel, Jacobi, Legendre, Weierstrasse, Hermite, Liouville, and Riemann. Unusual fullness of treatment, plus applications as well as theory, in discussing elliptic function (the universe of elliptic integrals originating in works of Abel and Jacobi), their existence, and ultimate meaning. Use is made of Riemann to provide the most general theory. 40 page table of formulas. 76 figures. xxiii + 498pp.
S483 Paperbound **$2.55**

THE THEORY AND FUNCTIONS OF A REAL VARIABLE AND THE THEORY OF FOURIER'S SERIES, E. W. Hobson. One of the best introductions to set theory and various aspects of functions and Fourier's series. Requires only a good background in calculus. Provides an exhaustive coverage of: metric and descriptive properties of sets of points; transfinite numbers and order types; functions of a real variable; the Riemann and Lebesgue integrals; sequences and series of numbers; power-series; functions representable by series sequences of continuous functions; trigonometrical series; representation of functions by Fourier's series; complete exposition (200pp.) on set theory; and much more. "The best possible guide," Nature. Vol. I: 88 detailed examples, 10 figures. Index. xv + 736pp. Vol. II: 117 detailed examples, 13 figures. Index. x + 780pp. 6⅛ x 9¼.
Vol. I: S387 Paperbound **$3.00**
Vol. II: S388 Paperbound **$3.00**

ALMOST PERIODIC FUNCTIONS, A. S. Besicovitch. This unique and important summary by a well-known mathematician covers in detail the two stages of development in Bohr's theory of almost periodic functions: (1) as a generalization of pure periodicity, with results and proofs; (2) the work done by Stepanoff, Wiener, Weyl, and Bohr in generalizing the theory. Bibliography. xi + 180pp. 5⅜ x 8.
S18 Paperbound **$1.75**

THE ANALYTICAL THEORY OF HEAT, Joseph Fourier. This book, which revolutionized mathematical physics, is listed in the Great Books program, and many other listings of great books. It has been used with profit by generations of mathematicians and physicists who are interested in either heat or in the application of the Fourier integral. Covers cause and reflection of rays of heat, radiant heating, heating of closed spaces, use of trigonometric series in the theory of heat, Fourier integral, etc. Translated by Alexander Freeman. 20 figures. xxii + 466pp. 5⅜ x 8.
S93 Paperbound **$2.50**

AN INTRODUCTION TO FOURIER METHODS AND THE LAPLACE TRANSFORMATION, Philip Franklin. Concentrates upon essentials, enabling the reader with only a working knowledge of calculus to gain an understanding of Fourier methods in a broad sense, suitable for most applications. This work covers complex qualities with methods of computing elementary functions for complex values of the argument and finding approximations by the use of charts; Fourier series and integrals with half-range and complex Fourier series; harmonic analysis; Fourier and Laplace transformations, etc.; partial differential equations with applications to transmission of electricity; etc. The methods developed are related to physical problems of heat flow, vibrations, electrical transmission, electromagnetic radiation, etc. 828 problems with answers. Formerly entitled "Fourier Methods." Bibliography. Index. x + 289pp. 5⅜ x 8.
S452 Paperbound **$2.00**

THE FOURIER INTEGRAL AND CERTAIN OF ITS APPLICATIONS, Norbert Wiener. The only book-length study of the Fourier integral as link between pure and applied math. An expansion of lectures given at Cambridge. Partial contents: Plancherel's theorem, general Tauberian theorem, special Tauberian theorems, generalized harmonic analysis. Bibliography. viii + 201pp. 5⅜ x 8.
S272 Paperbound **$1.50**

INTRODUCTION TO THE THEORY OF FOURIER'S SERIES AND INTEGRALS, H. S. Carslaw. 3rd revised edition. This excellent introduction is an outgrowth of the author's courses at Cambridge. Historical introduction, rational and irrational numbers, infinite sequences and series, functions of a single variable, definite integral, Fourier series, Fourier integrals, and similar topics. Appendixes discuss practical harmonic analysis, periodogram analysis. Lebesgue's theory. Indexes. 84 examples, bibliography. xii + 368pp. 5⅜ x 8. S48 Paperbound **$2.25**

FOURIER'S SERIES AND SPHERICAL HARMONICS, W. E. Byerly. Continues to be recognized as one of most practical, useful expositions. Functions, series, and their differential equations are concretely explained in great detail; theory is applied constantly to practical problems, which are fully and lucidly worked out. Appendix includes 6 tables of surface zonal harmonics, hyperbolic functions, Bessel's functions. Bibliography. 190 problems, approximately half with answers. ix + 287pp. 5⅜ x 8. S536 Paperbound **$1.75**

INFINITE SEQUENCES AND SERIES, Konrad Knopp. First publication in any language! Excellent introduction to 2 topics of modern mathematics, designed to give the student background to penetrate farther by himself. Sequences & sets, real & complex numbers, etc. Functions of a real & complex variable. Sequences & series. Infinite series. Convergent power series. Expansion of elementary functions. Numerical evaluation of series. Bibliography. v + 186pp. 5⅜ x 8. S153 Paperbound **$1.75**

TRIGONOMETRICAL SERIES, Antoni Zygmund. Unique in any language on modern advanced level. Contains carefully organized analyses of trigonometric, orthogonal, Fourier systems of functions, with clear adequate descriptions of summability of Fourier series, proximation theory, conjugate series, convergence, divergence of Fourier series. Especially valuable for Russian, Eastern European coverage. Bibliography. 329pp. 5⅜ x 8. S290 Paperbound **$2.00**

DICTIONARY OF CONFORMAL REPRESENTATIONS, H. Kober. Laplace's equation in 2 dimensions solved in this unique book developed by the British Admiralty. Scores of geometrical forms & their transformations for electrical engineers, Joukowski aerofoil for aerodynamists. Schwarz-Christoffel transformations for hydrodynamics, transcendental functions. Contents classified according to analytical functions describing transformation. Twin diagrams show curves of most transformations with corresponding regions. Glossary. Topological index. 447 diagrams. 244pp. 6⅛ x 9¼. S160 Paperbound **$2.00**

CALCULUS OF VARIATIONS, A. R. Forsyth. Methods, solutions, rather than determination of weakest valid hypotheses. Over 150 examples completely worked-out show use of Euler, Legendre, Jacobi, Weierstrass tests for maxima, minima. Integrals with one original dependent variable; derivatives of 2nd order; two dependent variables, one independent variable; double integrals involving 1 dependent variable, 2 first derivatives; double integrals involving partial derivatives of 2nd order; triple integrals; much more. 50 diagrams. 678pp. 5⅝ x 8⅜. S622 Paperbound **$2.95**

LECTURES ON THE CALCULUS OF VARIATIONS, O. Bolza. Analyzes in detail the fundamental concepts of the calculus of variations, as developed from Euler to Hilbert, with sharp formulations of the problems and rigorous demonstrations of their solutions. More than a score of solved examples; systematic references for each theorem. Covers the necessary and sufficient conditions; the contributions made by Euler, Du Bois-Reymond, Hilbert, Weierstrass, Legendre, Jacobi, Erdmann, Kneser, and Gauss; and much more. Index. Bibliography. xi + 271pp. 5⅜ x 8. S218 Paperbound **$1.65**

A TREATISE ON THE CALCULUS OF FINITE DIFFERENCES, G. Boole. A classic in the literature of the calculus. Thorough, clear discussion of basic principles, theorems, methods. Covers MacLaurin's and Herschel's theorems, mechanical quadrature, factorials, periodical constants, Bernoulli's numbers, difference-equations (linear, mixed, and partial), etc. Stresses analogies with differential calculus. 236 problems, answers to the numerical ones. viii + 336pp. 5⅜ x 8. S695 Paperbound **$1.85**

Prices subject to change without notice.

Dover publishes books on art, music, philosophy, literature, languages, history, social sciences, psychology, handcrafts, orientalia, puzzles and entertainments, chess, pets and gardens, books explaining science, intermediate and higher mathematics, mathematical physics, engineering, biological sciences, earth sciences, classics of science, etc. Write to:

Dept. catrr.
Dover Publications, Inc.
180 Varick Street, N.Y. 14, N.Y.

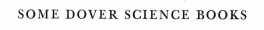

SOME DOVER SCIENCE BOOKS

SOME DOVER SCIENCE BOOKS

WHAT IS SCIENCE?,
Norman Campbell
This excellent introduction explains scientific method, role of mathematics, types of scientific laws. Contents: 2 aspects of science, science & nature, laws of science, discovery of laws, explanation of laws, measurement & numerical laws, applications of science. 192pp. 5⅜ x 8.   60043-2 Paperbound $1.25

FADS AND FALLACIES IN THE NAME OF SCIENCE,
Martin Gardner
Examines various cults, quack systems, frauds, delusions which at various times have masqueraded as science. Accounts of hollow-earth fanatics like Symmes; Velikovsky and. wandering planets; Hoerbiger; Bellamy and the theory of multiple moons; Charles Fort; dowsing, pseudoscientific methods for finding water, ores, oil. Sections on naturopathy, iridiagnosis, zone therapy, food fads, etc. Analytical accounts of Wilhelm Reich and orgone sex energy; L. Ron Hubbard and Dianetics; A. Korzybski and General Semantics; many others. Brought up to date to include Bridey Murphy, others. Not just a collection of anecdotes, but a fair, reasoned appraisal of eccentric theory. Formerly titled *In the Name of Science.* Preface. Index. x + 384pp. 5⅜ x 8.
20394-8 Paperbound $2.00

PHYSICS, THE PIONEER SCIENCE,
L. W. Taylor
First thorough text to place all important physical phenomena in cultural-historical framework; remains best work of its kind. Exposition of physical laws, theories developed chronologically, with great historical, illustrative experiments diagrammed, described, worked out mathematically. Excellent physics text for self-study as well as class work. Vol. 1: Heat, Sound: motion, acceleration, gravitation, conservation of energy, heat engines, rotation, heat, mechanical energy, etc. 211 illus. 407pp. 5⅜ x 8. Vol. 2: Light, Electricity: images, lenses, prisms, magnetism, Ohm's law, dynamos, telegraph, quantum theory, decline of mechanical view of nature, etc. Bibliography. 13 table appendix. Index. 551 illus. 2 color plates. 508pp. 5⅜ x 8.
60565-5, 60566-3 Two volume set, paperbound $5.50

THE EVOLUTION OF SCIENTIFIC THOUGHT FROM NEWTON TO EINSTEIN,
A. d'Abro
Einstein's special and general theories of relativity, with their historical implications, are analyzed in non-technical terms. Excellent accounts of the contributions of Newton, Riemann, Weyl, Planck, Eddington, Maxwell, Lorentz and others are treated in terms of space and time, equations of electromagnetics, finiteness of the universe, methodology of science. 21 diagrams. 482pp. 5⅜ x 8.
20002-7 Paperbound $2.50

CHANCE, LUCK AND STATISTICS: THE SCIENCE OF CHANCE,
Horace C. Levinson
Theory of probability and science of statistics in simple, non-technical language.
Part I deals with theory of probability, covering odd superstitions in regard to
"luck," the meaning of betting odds, the law of mathematical expectation,
gambling, and applications in poker, roulette, lotteries, dice, bridge, and other
games of chance. Part II discusses the misuse of statistics, the concept of statistical probabilities, normal and skew frequency distributions, and statistics applied to various fields—birth rates, stock speculation, insurance rates, advertis-
ing, etc. "Presented in an easy humorous style which I consider the best kind of
expository writing," Prof. A. C. Cohen, Industry Quality Control. Enlarged
revised edition. Formerly titled *The Science of Chance*. Preface and two new
appendices by the author. xiv + 365pp. 5⅜ x 8.   21007-3 Paperbound $2.00

BASIC ELECTRONICS,
prepared by the U.S. Navy Training Publications Center
A thorough and comprehensive manual on the fundamentals of electronics.
Written clearly, it is equally useful for self-study or course work for those with
a knowledge of the principles of basic electricity. Partial contents: Operating
Principles of the Electron Tube; Introduction to Transistors; Power Supplies
for Electronic Equipment; Tuned Circuits; Electron-Tube Amplifiers; Audio
Power Amplifiers; Oscillators; Transmitters; Transmission Lines; Antennas and
Propagation; Introduction to Computers; and related topics. Appendix. Index.
Hundreds of illustrations and diagrams. vi + 471pp. 6½ x 9¼.
61076-4 Paperbound $2.95

BASIC THEORY AND APPLICATION OF TRANSISTORS,
prepared by the U.S. Department of the Army
An introductory manual prepared for an army training program. One of the
finest available surveys of theory and application of transistor design and
operation. Minimal knowledge of physics and theory of electron tubes required.
Suitable for textbook use, course supplement, or home study. Chapters: Introduction; fundamental theory of transistors; transistor amplifier fundamentals;
parameters, equivalent circuits, and characteristic curves; bias stabilization;
transistor analysis and comparison using characteristic curves and charts; audio
amplifiers; tuned amplifiers; wide-band amplifiers; oscillators; pulse and switch-
ing circuits; modulation, mixing, and demodulation; and additional semi-
conductor devices. Unabridged, corrected edition. 240 schematic drawings,
photographs, wiring diagrams, etc. 2 Appendices. Glossary. Index. 263pp.
6½ x 9¼.   60380-6 Paperbound $1.75

GUIDE TO THE LITERATURE OF MATHEMATICS AND PHYSICS,
N. G. Parke III
Over 5000 entries included under approximately 120 major subject headings of
selected most important books, monographs, periodicals, articles in English,
plus important works in German, French, Italian, Spanish, Russian (many
recently available works). Covers every branch of physics, math, related engineering. Includes author, title, edition, publisher, place, date, number of
volumes, number of pages. A 40-page introduction on the basic problems of
research and study provides useful information on the organization and use of
libraries, the psychology of learning, etc. This reference work will save you
hours of time. 2nd revised edition. Indices of authors, subjects, 464pp. 5⅜ x 8.
60447-0 Paperbound $2.75

THE RISE OF THE NEW PHYSICS (formerly THE DECLINE OF MECHANISM), *A. d'Abro*
This authoritative and comprehensive 2-volume exposition is unique in scientific publishing. Written for intelligent readers not familiar with higher mathematics, it is the only thorough explanation in non-technical language of modern mathematical-physical theory. Combining both history and exposition, it ranges from classical Newtonian concepts up through the electronic theories of Dirac and Heisenberg, the statistical mechanics of Fermi, and Einstein's relativity theories. "A must for anyone doing serious study in the physical sciences," *J. of Franklin Inst.* 97 illustrations. 991pp. 2 volumes.

20003-5, 20004-3 Two volume set, paperbound $5.50

THE STRANGE STORY OF THE QUANTUM, AN ACCOUNT FOR THE GENERAL READER OF THE GROWTH OF IDEAS UNDERLYING OUR PRESENT ATOMIC KNOWLEDGE, *B. Hoffmann*
Presents lucidly and expertly, with barest amount of mathematics, the problems and theories which led to modern quantum physics. Dr. Hoffmann begins with the closing years of the 19th century, when certain trifling discrepancies were noticed, and with illuminating analogies and examples takes you through the brilliant concepts of Planck, Einstein, Pauli, de Broglie, Bohr, Schroedinger, Heisenberg, Dirac, Sommerfeld, Feynman, etc. This edition includes a new, long postscript carrying the story through 1958. "Of the books attempting an account of the history and contents of our modern atomic physics which have come to my attention, this is the best," H. Margenau, Yale University, in *American Journal of Physics.* 32 tables and line illustrations. Index. 275pp. 5⅜ x 8.

20518-5 Paperbound $2.00

GREAT IDEAS AND THEORIES OF MODERN COSMOLOGY, *Jagjit Singh*
The theories of Jeans, Eddington, Milne, Kant, Bondi, Gold, Newton, Einstein, Gamow, Hoyle, Dirac, Kuiper, Hubble, Weizsäcker and many others on such cosmological questions as the origin of the universe, space and time, planet formation, "continuous creation," the birth, life, and death of the stars, the origin of the galaxies, etc. By the author of the popular *Great Ideas of Modern Mathematics.* A gifted popularizer of science, he makes the most difficult abstractions crystal-clear even to the most non-mathematical reader. Index. xii + 276pp. 5⅜ x 8½.

20925-3 Paperbound $2.50

GREAT IDEAS OF MODERN MATHEMATICS: THEIR NATURE AND USE, *Jagjit Singh*
Reader with only high school math will understand main mathematical ideas of modern physics, astronomy, genetics, psychology, evolution, etc., better than many who use them as tools, but comprehend little of their basic structure. Author uses his wide knowledge of non-mathematical fields in brilliant exposition of differential equations, matrices, group theory, logic, statistics, problems of mathematical foundations, imaginary numbers, vectors, etc. Original publications, appendices. indexes. 65 illustr. 322pp. 5⅜ x 8. 20587-8 Paperbound $2.25

THE MATHEMATICS OF GREAT AMATEURS, *Julian L. Coolidge*
Great discoveries made by poets, theologians, philosophers, artists and other non-mathematicians: Omar Khayyam, Leonardo da Vinci, Albrecht Dürer, John Napier, Pascal, Diderot, Bolzano, etc. Surprising accounts of what can result from a non-professional preoccupation with the oldest of sciences. 56 figures. viii + 211pp. 5⅜ x 8½. 61009-8 Paperbound $2.00

COLLEGE ALGEBRA, *H. B. Fine*
Standard college text that gives a systematic and deductive structure to algebra; comprehensive, connected, with emphasis on theory. Discusses the commutative, associative, and distributive laws of number in unusual detail, and goes on with undetermined coefficients, quadratic equations, progressions, logarithms, permutations, probability, power series, and much more. Still most valuable elementary-intermediate text on the science and structure of algebra. Index. 1560 problems, all with answers. x + 631pp. 5⅜ x 8. 60211-7 Paperbound $2.75

HIGHER MATHEMATICS FOR STUDENTS OF CHEMISTRY AND PHYSICS, *J. W. Mellor*
Not abstract, but practical, building its problems out of familiar laboratory material, this covers differential calculus, coordinate, analytical geometry, functions, integral calculus, infinite series, numerical equations, differential equations, Fourier's theorem, probability, theory of errors, calculus of variations, determinants. "If the reader is not familiar with this book, it will repay him to examine it," *Chem. & Engineering News.* 800 problems. 189 figures. Bibliography. xxi + 641pp. 5⅜ x 8. 60193-5 Paperbound $3.50

TRIGONOMETRY REFRESHER FOR TECHNICAL MEN, *A. A. Klaf*
A modern question and answer text on plane and spherical trigonometry. Part I covers plane trigonometry: angles, quadrants, trigonometrical functions, graphical representation, interpolation, equations, logarithms, solution of triangles, slide rules, etc. Part II discusses applications to navigation, surveying, elasticity, architecture, and engineering. Small angles, periodic functions, vectors, polar coordinates, De Moivre's theorem, fully covered. Part III is devoted to spherical trigonometry and the solution of spherical triangles, with applications to terrestrial and astronomical problems. Special time-savers for numerical calculation. 913 questions answered for you! 1738 problems; answers to odd numbers. 494 figures. 14 pages of functions, formulae. Index. x + 629pp. 5⅜ x 8.
20371-9 Paperbound $3.00

CALCULUS REFRESHER FOR TECHNICAL MEN, *A. A. Klaf*
Not an ordinary textbook but a unique refresher for engineers, technicians, and students. An examination of the most important aspects of differential and integral calculus by means of 756 key questions. Part I covers simple differential calculus: constants, variables, functions, increments, derivatives, logarithms, curvature, etc. Part II treats fundamental concepts of integration: inspection, substitution, transformation, reduction, areas and volumes, mean value, successive and partial integration, double and triple integration. Stresses practical aspects! A 50 page section gives applications to civil and nautical engineering, electricity, stress and strain, elasticity, industrial engineering, and similar fields. 756 questions answered. 556 problems; solutions to odd numbers. 36 pages of constants, formulae. Index. v + 431pp. 5⅜ x 8. 20370-0 Paperbound $2.25

INTRODUCTION TO THE THEORY OF GROUPS OF FINITE ORDER, *R. Carmichael*
Examines fundamental theorems and their application. Beginning with sets, systems, permutations, etc., it progresses in easy stages through important types of groups: Abelian, prime power, permutation, etc. Except 1 chapter where matrices are desirable, no higher math needed. 783 exercises, problems. Index. xvi + 447pp. 5⅜ x 8. 60300-8 Paperbound $3.00

FIVE VOLUME "THEORY OF FUNCTIONS" SET BY KONRAD KNOPP

This five-volume set, prepared by Konrad Knopp, provides a complete and readily followed account of theory of functions. Proofs are given concisely, yet without sacrifice of completeness or rigor. These volumes are used as texts by such universities as M.I.T., University of Chicago, N. Y. City College, and many others. "Excellent introduction . . . remarkably readable, concise, clear, rigorous," *Journal of the American Statistical Association.*

ELEMENTS OF THE THEORY OF FUNCTIONS,
Konrad Knopp
This book provides the student with background for further volumes in this set, or texts on a similar level. Partial contents: foundations, system of complex numbers and the Gaussian plane of numbers, Riemann sphere of numbers, mapping by linear functions, normal forms, the logarithm, the cyclometric functions and binomial series. "Not only for the young student, but also for the student who knows all about what is in it," *Mathematical Journal.* Bibliography. Index. 140pp. 5⅜ x 8.   60154-4 Paperbound $1.50

THEORY OF FUNCTIONS, PART I,
Konrad Knopp
With volume II, this book provides coverage of basic concepts and theorems. Partial contents: numbers and points, functions of a complex variable, integral of a continuous function, Cauchy's integral theorem, Cauchy's integral formulae, series with variable terms, expansion of analytic functions in power series, analytic continuation and complete definition of analytic functions, entire transcendental functions, Laurent expansion, types of singularities. Bibliography. Index. vii + 146pp. 5⅜ x 8.   60156-0 Paperbound $1.50

THEORY OF FUNCTIONS, PART II,
Konrad Knopp
Application and further development of general theory, special topics. Single valued functions. Entire, Weierstrass, Meromorphic functions. Riemann surfaces. Algebraic functions. Analytical configuration, Riemann surface. Bibliography. Index. x + 150pp. 5⅜ x 8.   60157-9 Paperbound $1.50

PROBLEM BOOK IN THE THEORY OF FUNCTIONS, VOLUME 1.
Konrad Knopp
Problems in elementary theory, for use with Knopp's *Theory of Functions,* or any other text, arranged according to increasing difficulty. Fundamental concepts, sequences of numbers and infinite series, complex variable, integral theorems, development in series, conformal mapping. 182 problems. Answers. viii + 126pp. 5⅜ x 8.   60158-7 Paperbound $1.50

PROBLEM BOOK IN THE THEORY OF FUNCTIONS, VOLUME 2,
Konrad Knopp
Advanced theory of functions, to be used either with Knopp's *Theory of Functions,* or any other comparable text. Singularities, entire & meromorphic functions, periodic, analytic, continuation, multiple-valued functions, Riemann surfaces, conformal mapping. Includes a section of additional elementary problems. "The difficult task of selecting from the immense material of the modern theory of functions the problems just within the reach of the beginner is here masterfully accomplished," *Am. Math. Soc.* Answers. 138pp. 5⅜ x 8.
60159-5 Paperbound $1.50

NUMERICAL SOLUTIONS OF DIFFERENTIAL EQUATIONS,
H. Levy & E. A. Baggott
Comprehensive collection of methods for solving ordinary differential equations
of first and higher order. All must pass 2 requirements: easy to grasp and
practical, more rapid than school methods. Partial contents: graphical integration of differential equations, graphical methods for detailed solution. Numerical solution. Simultaneous equations and equations of 2nd and higher orders.
"Should be in the hands of all in research in applied mathematics, teaching,"
Nature. 21 figures. viii + 238pp. 5⅜ x 8.   60168-4 Paperbound $1.85

ELEMENTARY STATISTICS, WITH APPLICATIONS IN MEDICINE AND THE
BIOLOGICAL SCIENCES, *F. E. Croxton*
A sound introduction to statistics for anyone in the physical sciences, assuming no prior acquaintance and requiring only a modest knowledge of math.
All basic formulas carefully explained and illustrated; all necessary reference
tables included. From basic terms and concepts, the study proceeds to frequency
distribution, linear, non-linear, and multiple correlation, skewness, kurtosis,
etc. A large section deals with reliability and significance of statistical methods.
Containing concrete examples from medicine and biology, this book will prove
unusually helpful to workers in those fields who increasingly must evaluate,
check, and interpret statistics. Formerly titled "Elementary Statistics with Applications in Medicine." 101 charts. 57 tables. 14 appendices. Index. vi +
376pp. 5⅜ x 8.   60506-X Paperbound $2.25

INTRODUCTION TO SYMBOLIC LOGIC,
S. Langer
No special knowledge of math required — probably the clearest book ever
written on symbolic logic, suitable for the layman, general scientist, and philos-
opher. You start with simple symbols and advance to a knowledge of the
Boole-Schroeder and Russell-Whitehead systems. Forms, logical structure, classes,
the calculus of propositions, logic of the syllogism, etc. are all covered. "One
of the clearest and simplest introductions," *Mathematics Gazette.* Second enlarged, revised edition. 368pp. 5⅜ x 8.   60164-1 Paperbound $2.25

A SHORT ACCOUNT OF THE HISTORY OF MATHEMATICS,
W. W. R. Ball
Most readable non-technical history of mathematics treats lives, discoveries of
every important figure from Egyptian, Phoenician, mathematicians to late 19th
century. Discusses schools of Ionia, Pythagoras, Athens, Cyzicus, Alexandria,
Byzantium, systems of numeration; primitive arithmetic; Middle Ages, Renais-
sance, including Arabs, Bacon, Regiomontanus, Tartaglia, Cardan, Stevinus,
Galileo, Kepler; modern mathematics of Descartes, Pascal, Wallis, Huygens,
Newton, Leibnitz, d'Alembert, Euler, Lambert, Laplace, Legendre, Gauss,
Hermite, Weierstrass, scores more. Index. 25 figures. 546pp. 5⅜ x 8.
                                                           20630-0 Paperbound $2.75

INTRODUCTION TO NONLINEAR DIFFERENTIAL AND INTEGRAL EQUATIONS,
Harold T. Davis
Aspects of the problem of nonlinear equations, transformations that lead to
equations solvable by classical means, results in special cases, and useful
generalizations. Thorough, but easily followed by mathematically sophisticated
reader who knows little about non-linear equations. 137 problems for student
to solve. xv + 566pp. 5⅜ x 8½.   60971-5 Paperbound $2.75

AN INTRODUCTION TO THE GEOMETRY OF N DIMENSIONS,
D. H. Y. Sommerville
An introduction presupposing no prior knowledge of the field, the only book
in English devoted exclusively to higher dimensional geometry. Discusses
fundamental ideas of incidence, parallelism, perpendicularity, angles between
linear space; enumerative geometry; analytical geometry from projective and
metric points of view; polytopes; elementary ideas in analysis situs; content of
hyper-spacial figures. Bibliography. Index. 60 diagrams. 196pp. 5⅜ x 8.
60494-2 Paperbound $1.50

ELEMENTARY CONCEPTS OF TOPOLOGY, *P. Alexandroff*
First English translation of the famous brief introduction to topology for the
beginner or for the mathematician not undertaking extensive study. This unusually useful intuitive approach deals primarily with the concepts of complex,
cycle, and homology, and is wholly consistent with current investigations.
Ranges from basic concepts of set-theoretic topology to the concept of Betti
groups. "Glowing example of harmony between intuition and thought," David
Hilbert. Translated by A. E. Farley. Introduction by D. Hilbert. Index. 25
figures. 73pp. 5⅜ x 8.   60747-X Paperbound $1.25

ELEMENTS OF NON-EUCLIDEAN GEOMETRY,
D. M. Y. Sommerville
Unique in proceeding step-by-step, in the manner of traditional geometry.
Enables the student with only a good knowledge of high school algebra and
geometry to grasp elementary hyperbolic, elliptic, analytic non-Euclidean geom-
etries; space curvature and its philosophical implications; theory of radical
axes; homothetic centres and systems of circles; parataxy and parallelism;
absolute measure; Gauss' proof of the defect area theorem; geodesic representation; much more, all with exceptional clarity. 126 problems at chapter endings
provide progressive practice and familiarity. 133 figures. Index. xvi + 274pp.
5⅜ x 8.   60460-8 Paperbound $2.00

INTRODUCTION TO THE THEORY OF NUMBERS, *L. E. Dickson*
Thorough, comprehensive approach with adequate coverage of classical literature, an introductory volume beginners can follow. Chapters on divisibility,
congruences, quadratic residues & reciprocity. Diophantine equations, etc. Full
treatment of binary quadratic forms without usual restriction to integral coefficients. Covers infinitude of primes, least residues. Fermat's theorem. Euler's
phi function, Legendre's symbol, Gauss's lemma, automorphs, reduced forms,
recent theorems of Thue & Siegel, many more. Much material not readily
available elsewhere. 239 problems. Index. I figure. viii + 183pp. 5⅜ x 8.
60342-3 Paperbound $1.75

MATHEMATICAL TABLES AND FORMULAS,
compiled by Robert D. Carmichael and Edwin R. Smith
Valuable collection for students, etc. Contains all tables necessary in college
algebra and trigonometry, such as five-place common logarithms, logarithmic
sines and tangents of small angles, logarithmic trigonometric functions, natural
trigonometric functions, four-place antilogarithms, tables for changing from
sexagesimal to circular and from circular to sexagesimal measure of angles, etc.
Also many tables and formulas not ordinarily accessible, including powers,
roots, and reciprocals, exponential and hyperbolic functions, ten-place logarithms of prime numbers, and formulas and theorems from analytical and
elementary geometry and from calculus. Explanatory introduction. viii +
269pp. 5⅜ x 8½.   60111-0 Paperbound $1.50

A Source Book in Mathematics,
D. E. Smith
Great discoveries in math, from Renaissance to end of 19th century, in English translation. Read announcements by Dedekind, Gauss, Delamain, Pascal, Fermat, Newton, Abel, Lobachevsky, Bolyai, Riemann, De Moivre, Legendre, Laplace, others of discoveries about imaginary numbers, number congruence, slide rule, equations, symbolism, cubic algebraic equations, non-Euclidean forms of geometry, calculus, function theory, quaternions, etc. Succinct selections from 125 different treatises, articles, most unavailable elsewhere in English. Each article preceded by biographical introduction. Vol. I: Fields of Number, Algebra. Index. 32 illus. 338pp. 5⅜ x 8. Vol. II: Fields of Geometry, Probability, Calculus, Functions, Quaternions. 83 illus. 432pp. 5⅜ x 8.

60552-3, 60553-1 Two volume set, paperbound $5.00

Foundations of Physics,
R. B. Lindsay & H. Margenau
Excellent bridge between semi-popular works & technical treatises. A discussion of methods of physical description, construction of theory; valuable for physicist with elementary calculus who is interested in ideas that give meaning to data, tools of modern physics. Contents include symbolism; mathematical equations; space & time foundations of mechanics; probability; physics & continua; electron theory; special & general relativity; quantum mechanics; causality. "Thorough and yet not overdetailed. Unreservedly recommended," *Nature* (London). Unabridged, corrected edition. List of recommended readings. 35 illustrations. xi + 537pp. 5⅜ x 8.

60377-6 Paperbound $3.50

Fundamental Formulas of Physics,
ed. by D. H. Menzel
High useful, full, inexpensive reference and study text, ranging from simple to highly sophisticated operations. Mathematics integrated into text—each chapter stands as short textbook of field represented. Vol. 1: Statistics, Physical Constants, Special Theory of Relativity, Hydrodynamics, Aerodynamics, Boundary Value Problems in Math, Physics, Viscosity, Electromagnetic Theory, etc. Vol. 2: Sound, Acoustics, Geometrical Optics, Electron Optics, High-Energy Phenomena, Magnetism, Biophysics, much more. Index. Total of 800pp. 5⅜ x 8.

60595-7, 60596-5 Two volume set, paperbound $4.75

Theoretical Physics,
A. S. Kompaneyets
One of the very few thorough studies of the subject in this price range. Provides advanced students with a comprehensive theoretical background. Especially strong on recent experimentation and developments in quantum theory. Contents: Mechanics (Generalized Coordinates, Lagrange's Equation, Collision of Particles, etc.), Electrodynamics (Vector Analysis, Maxwell's equations, Transmission of Signals, Theory of Relativity, etc.), Quantum Mechanics (the Inadequacy of Classical Mechanics, the Wave Equation, Motion in a Central Field, Quantum Theory of Radiation, Quantum Theories of Dispersion and Scattering, etc.), and Statistical Physics (Equilibrium Distribution of Molecules in an Ideal Gas, Boltzmann Statistics, Bose and Fermi Distribution. Thermodynamic Quantities, etc.). Revised to 1961. Translated by George Yankovsky, authorized by Kompaneyets. 137 exercises. 56 figures. 529pp. 5⅜ x 8½.

60972-3 Paperbound $3.50

MATHEMATICAL PHYSICS, *D. H. Menzel*
Thorough one-volume treatment of the mathematical techniques vital for
classical mechanics, electromagnetic theory, quantum theory, and relativity.
Written by the Harvard Professor of Astrophysics for junior, senior, and graduate courses, it gives clear explanations of all those aspects of function theory,
vectors, matrices, dyadics, tensors, partial differential equations, etc., necessary
for the understanding of the various physical theories. Electron theory, relativity, and other topics seldom presented appear here in considerable detail.
Scores of definition, conversion factors, dimensional constants, etc. "More
detailed than normal for an advanced text . . . excellent set of sections on
Dyadics, Matrices, and Tensors," *Journal of the Franklin Institute*. Index. 193
problems, with answers. x + 412pp. 5⅜ x 8.   60056-4 Paperbound $2.50

THE THEORY OF SOUND, *Lord Rayleigh*
Most vibrating systems likely to be encountered in practice can be tackled
successfully by the methods set forth by the great Nobel laureate, Lord
Rayleigh. Complete coverage of experimental, mathematical aspects of sound
theory. Partial contents: Harmonic motions, vibrating systems in general, lateral
vibrations of bars, curved plates or shells, applications of Laplace's functions to
acoustical problems, fluid friction, plane vortex-sheet, vibrations of solid bodies,
etc. This is the first inexpensive edition of this great reference and study work.
Bibliography, Historical introduction by R. B. Lindsay. Total of 1040pp. 97
figures. 5⅜ x 8.   60292-3, 60293-1 Two volume set, paperbound $6.00

HYDRODYNAMICS, *Horace Lamb*
Internationally famous complete coverage of standard reference work on
dynamics of liquids & gases. Fundamental theorems, equations, methods, solutions, background, for classical hydrodynamics. Chapters include Equations of
Motion, Integration of Equations in Special Gases, Irrotational Motion, Motion
of Liquid in 2 Dimensions, Motion of Solids through Liquid-Dynamical Theory,
Vortex Motion, Tidal Waves, Surface Waves, Waves of Expansion, Viscosity,
Rotating Masses of Liquids. Excellently planned, arranged; clear, lucid presentation. 6th enlarged, revised edition. Index. Over 900 footnotes, mostly bibliographical. 119 figures. xv + 738pp. 6⅛ x 9¼.   60256-7 Paperbound $4.00

DYNAMICAL THEORY OF GASES, *James Jeans*
Divided into mathematical and physical chapters for the convenience of those
not expert in mathematics, this volume discusses the mathematical theory of
gas in a steady state, thermodynamics, Boltzmann and Maxwell, kinetic theory,
quantum theory, exponentials, etc. 4th enlarged edition, with new material on
quantum theory, quantum dynamics, etc. Indexes. 28 figures. 444pp. 6⅛ x 9¼.
        60136-6 Paperbound $2.75

THERMODYNAMICS, *Enrico Fermi*
Unabridged reproduction of 1937 edition. Elementary in treatment; remarkable
for clarity, organization. Requires no knowledge of advanced math beyond
calculus, only familiarity with fundamentals of thermometry, calorimetry.
Partial Contents: Thermodynamic systems; First & Second laws of thermodynamics; Entropy; Thermodynamic potentials: phase rule, reversible electric
cell; Gaseous reactions: van't Hoff reaction box, principle of LeChatelier;
Thermodynamics of dilute solutions: osmotic & vapor pressures, boiling &
freezing points; Entropy constant. Index. 25 problems. 24 illustrations. x +
160pp. 5⅜ x 8.   60361-X Paperbound $2.00

CELESTIAL OBJECTS FOR COMMON TELESCOPES,
Rev. T. W. Webb
Classic handbook for the use and pleasure of the amateur astronomer. Of inestimable aid in locating and identifying thousands of celestial objects. Vol I, The Solar System: discussions of the principle and operation of the telescope, procedures of observations and telescope-photography, spectroscopy, etc., precise location information of sun, moon, planets, meteors. Vol. II, The Stars: alphabetical listing of constellations, information on double stars, clusters, stars with unusual spectra, variables, and nebulae, etc. Nearly 4,000 objects noted. Edited and extensively revised by Margaret W. Mayall, director of the American Assn. of Variable Star Observers. New Index by Mrs. Mayall giving the location of all objects mentioned in the text for Epoch 2000. New Precession Table added. New appendices on the planetary satellites, constellation names and abbreviations, and solar system data. Total of 46 illustrations. Total of xxxix + 606pp. 5⅜ x 8. 20917-2, 20918-0 Two volume set, paperbound $5.00

PLANETARY THEORY,
E. W. Brown and C. A. Shook
Provides a clear presentation of basic methods for calculating planetary orbits for today's astronomer. Begins with a careful exposition of specialized mathematical topics essential for handling perturbation theory and then goes on to indicate how most of the previous methods reduce ultimately to two general calculation methods: obtaining expressions either for the coordinates of planetary positions or for the elements which determine the perturbed paths. An example of each is given and worked in detail. Corrected edition. Preface. Appendix. Index. xii + 302pp. 5⅜ x 8½. 61133-7 Paperbound $2.25

STAR NAMES AND THEIR MEANINGS,
Richard Hinckley Allen
An unusual book documenting the various attributions of names to the individual stars over the centuries. Here is a treasure-house of information on a topic not normally delved into even by professional astronomers; provides a fascinating background to the stars in folk-lore, literary references, ancient writings, star catalogs and maps over the centuries. Constellation-by-constellation analysis covers hundreds of stars and other asterisms, including the Pleiades, Hyades, Andromedan Nebula, etc. Introduction. Indices. List of authors and authorities. xx + 563pp. 5⅜ x 8½. 21079-0 Paperbound $3.00

A SHORT HISTORY OF ASTRONOMY, *A. Berry*
Popular standard work for over 50 years, this thorough and accurate volume covers the science from primitive times to the end of the 19th century. After the Greeks and the Middle Ages, individual chapters analyze Copernicus, Brahe, Galileo, Kepler, and Newton, and the mixed reception of their discoveries. Post-Newtonian achievements are then discussed in unusual detail: Halley, Bradley, Lagrange, Laplace, Herschel, Bessel, etc. 2 Indexes. 104 illustrations, 9 portraits. xxxi + 440pp. 5⅜ x 8. 20210-0 Paperbound $2.75

SOME THEORY OF SAMPLING, *W. E. Deming*
The purpose of this book is to make sampling techniques understandable to and useable by social scientists, industrial managers, and natural scientists who are finding statistics increasingly part of their work. Over 200 exercises, plus dozens of actual applications. 61 tables. 90 figs. xix + 602pp. 5⅜ x 8½.
61755-6 Paperbound $3.50

PRINCIPLES OF STRATIGRAPHY,
A. W. Grabau
Classic of 20th century geology, unmatched in scope and comprehensiveness. Nearly 600 pages cover the structure and origins of every kind of sedimentary, hydrogenic, oceanic, pyroclastic, atmoclastic, hydroclastic, marine hydroclastic, and bioclastic rock; metamorphism; erosion; etc. Includes also the constitution of the atmosphere; morphology of oceans, rivers, glaciers; volcanic activities; faults and earthquakes; and fundamental principles of paleontology (nearly 200 pages). New introduction by Prof. M. Kay, Columbia U. 1277 bibliographical entries. 264 diagrams. Tables, maps, etc. Two volume set. Total of xxxii + 1185pp. 5⅜ x 8.   60686-4, 60687-2 Two volume set, paperbound $6.25

SNOW CRYSTALS, *W. A. Bentley and W. J. Humphreys*
Over 200 pages of Bentley's famous microphotographs of snow flakes—the product of painstaking, methodical work at his Jericho, Vermont studio. The pictures, which also include plates of frost, glaze and dew on vegetation, spider webs, windowpanes; sleet; graupel or soft hail, were chosen both for their scientific interest and their aesthetic qualities. The wonder of nature's diversity is exhibited in the intricate, beautiful patterns of the snow flakes. Introductory text by W. J. Humphreys. Selected bibliography. 2,453 illustrations. 224pp. 8 x 10¼.   20287-9 Paperbound $3.25

THE BIRTH AND DEVELOPMENT OF THE GEOLOGICAL SCIENCES,
F. D. Adams
Most thorough history of the earth sciences ever written. Geological thought from earliest times to the end of the 19th century, covering over 300 early thinkers & systems: fossils & their explanation, vulcanists vs. neptunists, figured stones & paleontology, generation of stones, dozens of similar topics. 91 illustrations, including medieval, renaissance woodcuts, etc. Index. 632 footnotes, mostly bibliographical. 511pp. 5⅜ x 8.   20005-1 Paperbound $2.75

ORGANIC CHEMISTRY, *F. C. Whitmore*
The entire subject of organic chemistry for the practicing chemist and the advanced student. Storehouse of facts, theories, processes found elsewhere only in specialized journals. Covers aliphatic compounds (500 pages on the properties and synthetic preparation of hydrocarbons, halides, proteins, ketones, etc.), alicyclic compounds, aromatic compounds, heterocyclic compounds, organophosphorus and organometallic compounds. Methods of synthetic preparation analyzed critically throughout. Includes much of biochemical interest. "The scope of this volume is astonishing," *Industrial and Engineering Chemistry*. 12,000-reference index. 2387-item bibliography. Total of x + 1005pp. 5⅜ x 8.   60700-3, 60701-1 Two volume set, paperbound $4.50

THE PHASE RULE AND ITS APPLICATION,
Alexander Findlay
Covering chemical phenomena of 1, 2, 3, 4, and multiple component systems, this "standard work on the subject" (*Nature*, London), has been completely revised and brought up to date by A. N. Campbell and N. O. Smith. Brand new material has been added on such matters as binary, tertiary liquid equilibria, solid solutions in ternary systems, quinary systems of salts and water. Completely revised to triangular coordinates in ternary systems, clarified graphic representation, solid models, etc. 9th revised edition. Author, subject indexes. 236 figures. 505 footnotes, mostly bibliographic. xii + 494pp. 5⅜ x 8.
60091-2 Paperbound $2.75

A Course in Mathematical Analysis,
Edouard Goursat

Trans. by E. R. Hedrick, O. Dunkel, H. G. Bergmann. Classic study of fundamental material thoroughly treated. Extremely lucid exposition of wide range of subject matter for student with one year of calculus. Vol. 1: Derivatives and differentials, definite integrals, expansions in series, applications to geometry. 52 figures, 556pp. 60554-X Paperbound $3.00. Vol. 2, Part I: Functions of a complex variable, conformal representations, doubly periodic functions, natural boundaries, etc. 38 figures, 269pp. 60555-8 Paperbound $2.25. Vol. 2, Part II: Differential equations, Cauchy-Lipschitz method, nonlinear differential equations, simultaneous equations, etc. 308pp. 60556-6 Paperbound $2.50. Vol. 3, Part I: Variation of solutions, partial differential equations of the second order. 15 figures, 339pp. 61176-0 Paperbound $3.00. Vol. 3, Part II: Integral equations, calculus of variations. 13 figures, 389pp. 61177-9 Paperbound $3.00 60554-X, 60555-8, 60556-6 61176-0, 61177-9 Six volume set,
paperbound $13.75

Planets, Stars and Galaxies,
A. E. Fanning

Descriptive astronomy for beginners: the solar system; neighboring galaxies; seasons; quasars; fly-by results from Mars, Venus, Moon; radio astronomy; etc. all simply explained. Revised up to 1966 by author and Prof. D. H. Menzel, former Director, Harvard College Observatory. 29 photos, 16 figures. 189pp. 5⅜ x 8½.
21680-2 Paperbound $1.50

Great Ideas in Information Theory, Language and Cybernetics,
Jagjit Singh

Winner of Unesco's Kalinga Prize covers language, metalanguages, analog and digital computers, neural systems, work of McCulloch, Pitts, von Neumann, Turing, other important topics. No advanced mathematics needed, yet a full discussion without compromise or distortion. 118 figures. ix + 338pp. 5⅜ x 8½.
21694-2 Paperbound $2.25

Geometric Exercises in Paper Folding,
T. Sundara Row

Regular polygons, circles and other curves can be folded or pricked on paper, then used to demonstrate geometric propositions, work out proofs, set up well-known problems. 89 illustrations, photographs of actually folded sheets. xii + 148pp. 5⅜ x 8½.
21594-6 Paperbound $1.00

Visual Illusions, Their Causes, Characteristics and Applications,
M. Luckiesh

The visual process, the structure of the eye, geometric, perspective illusions, influence of angles, illusions of depth and distance, color illusions, lighting effects, illusions in nature, special uses in painting, decoration, architecture, magic, camouflage. New introduction by W. H. Ittleson covers modern developments in this area. 100 illustrations. xxi + 252pp. 5⅜ x 8.
21530-X Paperbound $1.50

Atoms and Molecules Simply Explained,
B. C. Saunders and R. E. D. Clark

Introduction to chemical phenomena and their applications: cohesion, particles, crystals, tailoring big molecules, chemist as architect, with applications in radioactivity, color photography, synthetics, biochemistry, polymers, and many other important areas. Non technical. 95 figures. x + 299pp. 5⅜ x 8½.
21282-3 Paperbound $1.50

THE PRINCIPLES OF ELECTROCHEMISTRY,
D. A. MacInnes

Basic equations for almost every subfield of electrochemistry from first principles, referring at all times to the soundest and most recent theories and results; unusually useful as text or as reference. Covers coulometers and Faraday's Law, electrolytic conductance, the Debye-Hueckel method for the theoretical calculation of activity coefficients, concentration cells, standard electrode potentials, thermodynamic ionization constants, pH, potentiometric titrations, irreversible phenomena. Planck's equation, and much more. 2 indices. Appendix. 585-item bibliography. 137 figures. 94 tables. ii + 478pp. 5⅝ x 8⅜.
60052-1 Paperbound $3.00

MATHEMATICS OF MODERN ENGINEERING,
E. G. Keller and R. E. Doherty

Written for the Advanced Course in Engineering of the General Electric Corporation, deals with the engineering use of determinants, tensors, the Heaviside operational calculus, dyadics, the calculus of variations, etc. Presents underlying principles fully, but emphasis is on the perennial engineering attack of set-up and solve. Indexes. Over 185 figures and tables. Hundreds of exercises, problems, and worked-out examples. References. Total of xxxiii + 623pp. 5⅜ x 8.
60734-8, 60735-6 Two volume set, paperbound $3.70

AERODYNAMIC THEORY: A GENERAL REVIEW OF PROGRESS,
William F. Durand, editor-in-chief

A monumental joint effort by the world's leading authorities prepared under a grant of the Guggenheim Fund for the Promotion of Aeronautics. Never equalled for breadth, depth, reliability. Contains discussions of special mathematical topics not usually taught in the engineering or technical courses. Also: an extended two-part treatise on Fluid Mechanics, discussions of aerodynamics of perfect fluids, analyses of experiments with wind tunnels, applied airfoil theory, the nonlifting system of the airplane, the air propeller, hydrodynamics of boats and floats, the aerodynamics of cooling, etc. Contributing experts include Munk, Giacomelli, Prandtl, Toussaint, Von Karman, Klemperer, among others. Unabridged republication. 6 volumes. Total of 1,012 figures, 12 plates, 2,186pp. Bibliographies. Notes. Indices. 5⅜ x 8½.   61709-2,
61710-6, 61711-4, 61712-2, 61713-0, 61715-9 Six volume set, paperbound $13.50

FUNDAMENTALS OF HYDRO- AND AEROMECHANICS,
L. Prandtl and O. G. Tietjens

The well-known standard work based upon Prandtl's lectures at Goettingen. Wherever possible hydrodynamics theory is referred to practical considerations in hydraulics, with the view of unifying theory and experience. Presentation is extremely clear and though primarily physical, mathematical proofs are rigorous and use vector analysis to a considerable extent. An Engineering Society Monograph, 1934. 186 figures. Index. xvi + 270pp. 5⅜ x 8.
60374-1 Paperbound $2.25

APPLIED HYDRO- AND AEROMECHANICS,
L. Prandtl and O. G. Tietjens

Presents for the most part methods which will be valuable to engineers. Covers flow in pipes, boundary layers, airfoil theory, entry conditions, turbulent flow in pipes, and the boundary layer, determining drag from measurements of pressure and velocity, etc. Unabridged, unaltered. An Engineering Society Monograph. 1934. Index. 226 figures, 28 photographic plates illustrating flow patterns. xvi + 311pp. 5⅜ x 8.   60375-X Paperbound $2.50

APPLIED OPTICS AND OPTICAL DESIGN,
A. E. Conrady
With publication of vol. 2, standard work for designers in optics is now complete for first time. Only work of its kind in English; only detailed work for practical designer and self-taught. Requires, for bulk of work, no math above trig. Step-by-step exposition, from fundamental concepts of geometrical, physical optics, to systematic study, design, of almost all types of optical systems. Vol. 1: all ordinary ray-tracing methods; primary aberrations; necessary higher aberration for design of telescopes, low-power microscopes, photographic equipment. Vol. 2: (Completed from author's notes by R. Kingslake, Dir. Optical Design, Eastman Kodak.) Special attention to high-power microscope, anastigmatic photographic objectives. "An indispensable work," *J., Optical Soc. of Amer.* Index. Bibliography. 193 diagrams. 852pp. 6⅛ x 9¼.
60611-2, 60612-0 Two volume set, paperbound $8.00

MECHANICS OF THE GYROSCOPE, THE DYNAMICS OF ROTATION,
R. F. Deimel, Professor of Mechanical Engineering at Stevens Institute of Technology
Elementary general treatment of dynamics of rotation, with special application of gyroscopic phenomena. No knowledge of vectors needed. Velocity of a moving curve, acceleration to a point, general equations of motion, gyroscopic horizon, free gyro, motion of discs, the damped gyro, 103 similar topics. Exercises. 75 figures. 208pp. 5⅜ x 8.
60066-1 Paperbound $1.75

STRENGTH OF MATERIALS,
J. P. Den Hartog
Full, clear treatment of elementary material (tension, torsion, bending, compound stresses, deflection of beams, etc.), plus much advanced material on engineering methods of great practical value: full treatment of the Mohr circle, lucid elementary discussions of the theory of the center of shear and the "Myosotis" method of calculating beam deflections, reinforced concrete, plastic deformations, photoelasticity, etc. In all sections, both general principles and concrete applications are given. Index. 186 figures (160 others in problem section). 350 problems, all with answers. List of formulas. viii + 323pp. 5⅜ x 8.
60755-0 Paperbound $2.50

HYDRAULIC TRANSIENTS,
G. R. Rich
The best text in hydraulics ever printed in English . . . by former Chief Design Engineer for T.V.A. Provides a transition from the basic differential equations of hydraulic transient theory to the arithmetic integration computation required by practicing engineers. Sections cover Water Hammer, Turbine Speed Regulation, Stability of Governing, Water-Hammer Pressures in Pump Discharge Lines, The Differential and Restricted Orifice Surge Tanks, The Normalized Surge Tank Charts of Calame and Gaden, Navigation Locks, Surges in Power Canals—Tidal Harmonics, etc. Revised and enlarged. Author's prefaces. Index. xiv + 409pp. 5⅜ x 8½.
60116-1 Paperbound $2.50

Prices subject to change without notice.

Available at your book dealer or write for free catalogue to Dept. Adsci, Dover Publications, Inc., 180 Varick St., N.Y., N.Y. 10014. Dover publishes more than 150 books each year on science, elementary and advanced mathematics, biology, music, art, literary history, social sciences and other areas.